The Five-Hundred-Year Plan

*A Town Primarily for People*

Integrated with Nature, for Today and the New Millennium

*by*

L. Gene Zellmer, AIA

Order this book online at www.trafford.com
or email orders@trafford.com

Most Trafford titles are also available at major online book retailers.

Print information available on the last page.

ISBN: 978-1-4120-1284-3 (sc)
ISBN: 978-1-4122-1731-6 (e)

*Trafford rev. 05/25/2021*

**North America & international**
toll-free: 844-688-6899 (USA & Canada)
fax: 812 355 4082

*Dedicated to*

*my*

*home-designing*

*Mother*

*and*

*home-building*

*Father.*

# Acknowledgments

To all of you who have been traveling along this rushing freeway of life over these many years within talking distance, who have shared your time and thoughts, have been contributors to this effort, ... to you I give my sincere gratitude. That includes those who simply made thought-provoking comments, but especially my staff, associates, partners, and the many great clients I have been so fortunate to work with.

Special gratitude is reserved for my family and particularly my wife, who has been patient, supportive, and encouraging. Her dedication of time and effort to improve the quality of my writing, to the degree it has been possible, has been very necessary and greatly appreciated.

Also, sincere appreciation to Cheryl Jencks, who has been my patient editor and valued advisor, and who prepared the final computer files for publication.

# Preface

This patchwork of essays hopes to present a new way for thinking about human habitats on Earth. It offers new possibilities for the paradigm shift that is needed.

The Table of Contents outlines the subtle differences in understanding what we now build. For example, the advantages of simply reinventing the individual home site triggered entirely new arrangements.

Part Four, Habitat Solutions, describes the modified spatial arrangements; but to be convinced of why or how they are necessary, it will be helpful to read the first parts of the book.

Many essays repeat relevant points, so it is possible to hop to different parts of particular interest - whatever makes it easier to get inside the idea. The thinking also builds from the early essays.

Ideally, it will be an adventure: discovering new spatial arrangements allowing for all the living hoped for in a town, from private moments alone with a great view to being surrounded by community life. The difficult part to imagine is reaching beyond the clumsiness and limitations of the words and drawings, to imagine the kind of perfection that often has come to historic towns only after centuries of refinement.

Many will be surprised that a three dimensional framework or armature can give people more freedom than our current 2-D framework spread out on the land.

None of it is entirely new, just new spatial relationship that will allow us to comprehensively meet the very serious challenges currently facing our towns and be sustainable for hundreds of years.

# Table of Contents

## *Part One: Vision of a Goal*

## *Part Two: Observations*

### *A: General Considerations*

## Part Four: Habitat Solutions

## Part Five: Habitat Support Systems

## Part Six: Habitat Sites

## *Part Seven: First Prototype?*

## *Appendix*

# Images

The images included throughout this book simply suggest a general impression and may have only an indirect connection with the text. Specifics of the images are usually not important to the points being made. When they are important or help clarify a point in the text, the connection will be mentioned or it will be obvious.

*Part One*

# Vision of a Goal

# Cities on Planet Earth

*Natural principles and towns*

Imagine you are from another planet and had never before visited Earth, but you had seen a box full of samples collected from it. Things like seashells, coral, sponges, animal skeletons, leaves, flowers, wasp nest, honey combs, bird's nest of straw and one of mud, all of which contained a consistent basic visual harmony of natural beauty. For the first time you are about to visit a city built by homo sapiens, the Earth's most advanced life form. Wouldn't you expect it to look as if it were based on the same natural principles of beauty and structure—that it blended into its earthly setting as comfortably as all those other things created by supposedly less-intelligent life forms?

And wouldn't you also expect it to function with the same kind of harmony, not causing any damage to its surroundings? That it borrowed resources from all around it for its own sustenance and then returned products that enhanced and nourished all the other surrounding life forms?

Perhaps you would expect it to look something like the picture on the cover of this book. The town concept shown provides the same number of homes as typically built in a square mile (640 acres) of suburbia. Instead of covering the land with streets and housing, however, this concept leaves 500 acres as open space. Every home has a back yard that's like being on a hill overlooking the open countryside, and each home also has a front porch on Main Street, with all the conveniences of urban life.

Why don't our cities look and function with the same natural harmony found in all other life forms? Unfortunately, we have usually thought of nature as the problem. We clear it out, scrape it clean, and build our boxes. We are knowledgeable, but probably not as smart as we think. When we notice problems, solutions tend to be band-aids, cover-it-over, patch-it-up, just enough to ameliorate the immediate concern.

We act like we live in a separate world. We've created totally man-made environments that appear foreign to most of the fundamental principles of nature. And as long as nature can tolerate our near-adolescent irresponsibility, it has let us continue in our naiveté and shrugged. Nature simply continues along trying to ignore or cover up our messes, ... trusting "survival of the fittest" as the rule. Perhaps

3

we can dominate nature, counteract every challenge it presents, beat it back, overpower it if necessary, and be the last remaining survivor in this process. Is that what we want, ... really? This planet offers us everything it has.

In terms of our cities, we've simply never accepted any responsibility beyond providing for ourselves. It seems to be a powerful but disconnected survival instinct. We seem to be the only life form with this disconnect. Why? Because we are the only ones with the freedom of that choice.

In many ways we are progressing to dramatically new levels of understanding, but we are also moving toward bigger populations and more self-absorbed confidence. If we can survive long enough, with enough sustained economic health to provide the extra funds necessary to finance developing appropriate technologies and gaining adequate wisdom, ... for the first time in all of history, we may arrive at the unique opportunity to fully accept our responsibility as a life form that can coexist as an integral part of our planet. It's almost within our reach.

How can human habitats do that? That's what the years of research described in this book have attempted to explore. It's as comprehensive as time has allowed, but the results should be considered a first step rather than a complete or final solution.

This concept simply attempts to pull together and arrange in new ways the basic down-to-earth solutions that have been known to work in the past for people's highest benefit and enjoyment. It simultaneously attempts to combine those solutions with the most advanced techniques for complementing the systems of nature.

It's not about inventing complex intellectual theories. With the help of others, however, some may emerge as our mutual progress evolves and the debate expands to higher levels of understanding, hopefully to help us more fully meet our responsibilities and potential.

# What's the Goal?
*Place for dreams … L.A.?*

A Place for Dreams
… with that special small-town atmosphere.
Resting or puttering in my garden
on the side of a hill overlooking
trees, lake, and distant farms,
a gentle breeze with a
hint of fresh-mown hay.

Could this be L.A.?

Watching neighbor kids with happy sounds
romping in our place for play,
watching friends at the local sidewalk cafe
greeting other town folk browsing
at shops along the way.

While I'm swinging,
with a wave to anyone
who happens to glance my way,
here on my front porch,

Could this be L.A.?

Yes, it may be, … someday.

If a town could be built that simply combined the few features mentioned in this brief verse for each individual home, it would revolutionize every aspect of family and community life. It would provide a first step in restoring many livable features of small towns that current towns have forfeited. No other existing or proposed town concept manages to combine these few basic features for every home. That's a big contrast from L.A. today.

For a climate as perfect as Southern California's, it's tragic to see what has been built there and what people put up with. When that vast, endless sea of housing and infrastructure gets too old to justify its maintenance, there may be a chance for something better.

Think of the best places you have been and appreciated, those that added to the best moments in your life, places that made you think they allowed a better way to live. Start your own list of the elements that contribute to the character of such places. Because you, your friends, and I all have the potential to totally reinvent the kind of towns where we and our grandchildren could live in the future.

We are at the threshold of potential change greater than any the world has ever seen. There are incredible new opportunities for increasing livability and sustainability. Our homes, towns, and major cities will be able to be built and function in a combination of ways never before imagined.

For the first time, cities will be free to be designed just for people. No compromises for services, transportation, or utilities, because all those can be provided without being noticed. Everything we need can be within a short walk.

The goal is to develop long-term planning objectives, including maximizing all imaginable possibilities, that will allow such a new concept town to fully and continually evolve.

Among the many possibilities there are two design objectives intended to give each home the best of two worlds. Each home will be able to have a front porch in the finest urban context and a back yard in the country.

# Did They Expect Big Old Cities?

*Big cities 200 years ago were discernible*

When those great centers of cities we so admire were first being built, no one ever imagined some of those cities would get as big as they are. Less than 200 years ago the extent of every big city could be seen from a single local high vantage point. They had a comprehensible scale.

Those grand piazzas and market halls of the Middle Ages were all about marketing. They were designed to impress people and function as places for trade. Many massive grand stone halls were built over outdoor open markets because this was accepted as public space. Tall wide arches allowed daylight as far back into the market structure as possible. Upper floors were for some municipal offices and storage of goods.

On certain days the large plaza spaces were gigantic parking lots for horses, wagons, and the carriages of wealthy buyers from many parts of the world. They were a busy mess, built for city life, parades, and gallantry, but primarily as the place for business, the *reason* for cities. These were the giant parking lots and big box stores of those days.

Today those same busy plazas are for strolling and lounging tourists admiring the old buildings and watching each other. Not entirely by accident, these are the hearts of our most livable cities, from which we have much to learn. Such places have matured and been reworked over hundreds of years.

But cities are still relatively new developments. Mark Girouard in *Cities & People* offers us some insight. As recently as 1900, even after the greatest romantic times of livable plazas, town centers, impressive buildings, great music, fancy dress, high-class manners, scientific answers for everything, and worldly philosophies, few cities had sewer systems. The perfume industry played a big role in the livability of some cities. Most toilet and waste systems were simply a chamber pot, the contents of which were dumped daily. If you were fortunate, the waste was hauled away in carriages and dumped in big piles at the edge of town.

It's no surprise that we have not yet even begun to envision human habitats that are truly integrated with nature.

Creating livability opportunities in cities is a more sophisticated challenge than our modern cities seem to be able to meet. After all, what is the primary purpose of cities? Modern cities are not made for people. They are, and may always have been, for what people make, do, or trade, plus transportation. Usually any enjoyment is secondary. (There's excellent information on addressing these issues from *Making Cities Livable*, www.liveablecities.org).

Nevertheless, people become attached to their hometowns or cities. Memories of personal or childhood experiences affect how they look at their surroundings. It's their concept of place, of what a town *is*. It can be anything, with conventional streets and houses; it can be one story or ten. People will ignore its negatives, even defend it—it's like a codependent connection to the past. This happens even with places obviously laid out primarily for cars instead of people.

Perhaps for the first time in history, we have the opportunity to completely turn this upside-down situation right-side up. Today we have the ability to make a town just for people, and make all those other things secondary. This making and doing what we currently call a city is primitive compared to alternatives we are beginning to recognize and understand. We can do better. Humans must assume our responsibility as the most advanced life form on earth. What we do here is not just about us.

We need to assume responsibility for making great and permanent cities that will last for centuries into the future. Our worth as ancestors will be measured by the value of our contribution to the long-term progress of humanity.

If we want a place to be recognized as the city center, as important as what remains of medieval centers, then it must be built just as permanently. If permanent, then it must be done and located correctly. That means comprehensively addressing housing, transportation, and all environmental issues. We need to expect that plans will be made for and carried out over many decades, even centuries, and to be flexible enough for refinement over time.

Most general plans talk about 20 or 50 years. It's a joke to think of those as plans for the future. Incremental improvements will never be able to meet the challenges we will face in the next few centuries. Many "improvements" will simply lock up the use of certain land parcels and keep them from being developed comprehensively at the appropriate time. The better the quality of construction, the longer they will last, and the longer the land will be locked up.

For decades into the future, *general plans that designate designs from the past as our primary options could become the greatest enemy of better, long-term, comprehensive planning concepts.*

We should demand and plan for growth, for more livable towns, for a different kind of big city, and for meeting all our responsibilities with permanent qualities and human scale.

We can start today with the great advantage of new perspectives of understanding and technologies that have never existed before.

# City Problems and Solutions

*A different twist and several new issues*

What's wrong with our current cities? There are many books on this subject, so there's no need to repeat that material here. Some are listed in the bibliography.

Many authors do a remarkable job describing what's wrong with cities and the challenges we face, usually in the first sections of their books. Everyone has seen concerns presented in various media about traffic, sprawl, lost farmland, affordability, and our lack of livable community spaces.

To understand current problems you may find it helpful to read Marshall, Register, and Kunstler. To expand your historic perspective, try Girouard and Southworth. The thoughts of Jane Jacobs, McHarg, Lynch, Gallion, Schneider, and Soleri were particularly influential on my initial interest in this subject. My sketches were used in a booklet Kenneth R. Schneider wrote in the 1970s called *The Community Space Frame*, but back then I hadn't yet worked out any architectural or planning solutions.

This search for an entirely new concept for towns recognizes and grows from the concerns everyone has presented, but it expands on them with a slightly different twist and adds a few issues not normally considered. Then, based on all that, this book concludes by describing a totally new kind of solution. Change for the purpose of change is *not* one of the objectives.

The solution proposed is a three-dimensional rearrangement of town elements to meet activity and space needs for people to have maximum livability. Any particular community and culture will determine the appearance. Any design character shown in the images is only to give scale.

After exploring some basic thinking affecting possible combinations of spatial relationships, it should become clear why a major paradigm shift in our concept of a town is necessary—a shift much more comprehensive than any others thus far proposed.

The other books also continue in their remaining sections with each author's proposed solutions. I generally agree with most of the objectives and the solutions for the circumstance any particular author intended to deal with.

However, the concept in this book recognizes and is generally directed toward solving a different and more comprehensive set of circumstances. Therefore, it's not in conflict with other proposed solutions. It especially has no conflict with those that address and solve immediate problems, solutions which many of our cities so desperately need. Where appropriate, this concept does borrow and apply some of the suggestions already made by others, but with very different sets of arrangements growing out a different understanding of the problems.

This question does exist, "Will the status quo support anything totally new?" Something totally new will affect everyone. The work they currently do will probably be changed.

Participants in the status quo or anything currently existing are intellectually, emotionally, or financially invested in it. For some, if it means change or doesn't follow familiar rules, it's not easy to see benefits beyond that. I hope they will be patient with the presentation of this concept until they have enough information to visualize living in it.

To freely develop a fresh concept, stereotyped opinions of design ideas based on any previous applications had to be ignored. From the beginning of this research effort, every option had to be considered in order to test for the best of every imaginable possible combination of solutions. Something that didn't work in the past may work in a new context, where each could be complemented by entirely new relationships and arrangements. Naturally the net results were expected to be different.

Initially, something totally new that affects everyone is likely to be opposed by everyone. It maybe helpful to remember that any change of this magnitude will be very gradual.

We have a better chance of meeting our current and future housing and community challenges if we're prepared to accept a paradigm shift in every element related to all the issues affecting human habitats.

# Paradigm Shift to Where?

*A comprehensive solution to satisfy all concerns*

Before making some great paradigm shift forward, consider the fact that we may need a couple of paradigm shifts at our roots.

The truly beautiful natural things in the world have been beautiful through all time and for all peoples. The same can be said for satisfying human-created places. Enduring examples have been satisfying throughout time, surviving different governments, religions, and economic times, ... satisfying for all people, no matter the race or nationality. Tourists travel all over the world to experience and enjoy such places.

However, it's important to remember that even beautiful Medieval cities had just as much exploitation and misery as any modern city.

Today there is considerable struggle between the different approaches for how to address the problems facing our cities. On the city and county level, issues often surround master plans and housing. Everyone wants a paradigm shift, as long as it's your thinking shifting to my thinking.

It seems that current efforts end up limiting their focus on problems and strategies. Rather than dealing with causes, treatment of the symptom is considered a solution. Those involved are often forced to accept a less-than-satisfactory compromise.

Any effort to solve a specific current condition is surrounded by a huge amount of complexity and bureaucracy. Even if an incremental improvement is made, all the surrounding baggage is usually dragged along with it.

There are preconceived approaches and hidden agendas that dominate many discussions. Usually no vision for a comprehensive solution is even being proposed. Sometimes the objective is simply to object.

Some feel, even without any real concept as an objective, that laws can coerce the correct objectives magically into existence, that government power is the only useful tool. Others feel practical solutions can only come from those close to the work, that materials and economics show the best way to meet people's needs. Lengthy debates end up being about procedures, laws, ordinances, control, power, race, and political philosophies.

These comments are not intended to trivialize such activities. Unfortunately, they seem to be the only options for dealing with our current basic design parameters. Entire careers become trapped in these exercises. It's the classic "can't see the forest for all the trees problem."

What's needed are entirely new basic parameters for the design of towns.

With such current states of mind, if a new comprehensive and perfect solution for a town was presented, preconceived thinking held by many would prevent them from even recognizing it. All sides' first reaction would be to oppose it.

The reality is that a truly comprehensive solution would satisfy all sides. Such a solution would provide an opportunity for all sets of talents to work together to solve our mutual growing challenges. This is what our future will demand. *This is the first, most important place the paradigm must shift.*

It may be impossible to get to where we need to be in our thinking by starting from where we are.

Another issue, mutual respect, seems to be one of the missing ingredients in our society. Many previous generations went to school with five grades and 48 kids in one classroom with one teacher. It would never have been possible without each individual's respect for the teacher and other classmates. Often there was not a common language, but for time spent in school the results were considerably better than they are today. We need to restore a similar attitude of mutual respect.

The challenges we face will require the best input possible from every source. Preconceived, prejudiced animosities among certain sectors of our society need to be forgotten. Reasons for different points of view need to be understood, appreciated, and combined in ways that complement each other. Nothing will be satisfied by partial solutions.

We need to better understand every aspect of how we live daily, … in our home, neighborhood, and town, and then to provide places and arrangements to enrich each of those. As the basics are pulled together, techniques can evolve for combining them with the best of existing and *historically important towns*, being careful not to compromise the important values of either.

We need to start with a blank piece of paper and a box full of the best experiences from the past and the best technologies for the future. By placing on the paper only combinations of what is needed *and* what will work together, the unnecessary will be left behind.

Unencumbered design solutions need to be discovered for every activity done by each individual in the normal pattern of living and all relationships with everything that makes up community life. This suggests a paradigm shift back to such basics.

# Durability Creates Value
*Permanent: more sustainable than recycling or reuse*

Stone buildings of the Middle Ages show us the value of permanent structures. The builders probably didn't even give that a thought. Stacking stones was simply the way they built useful buildings. After all, in some cases they were taking the stones from other equally important structures built hundreds of years before, buildings that no longer suited their needs or their culture. The structures they disassembled or remodeled must have given them some clue that buildings can last a long, long time. But then, civic or commercial structures were simply built for some guy or some town with enough money, someone who was wanting to make more by appearing substantial and providing needed spaces in order to attract buyers, sellers, and even tourists.

They certainly had no idea they would teach us that *permanent buildings are more sustainable* than recycling or even reuse of materials.

But we have some problems to overcome. To get livability we think we can fake it simply by building the equivalent of Hollywood stage sets for our towns, mimicking the shapes and surfaces of the past. If those old-timers had had sheet rock and foam board with ¹⁄₁₆″ epoxy coatings, they probably would have used them to build those great market halls and cathedrals. Today there would be nothing left, and a large percentage of Europeans wouldn't have those centuries-old stone structures to live in. The housing shortage would be worse than it is.

Permanence is not automatic in our time. It requires conscious decision and foresight: the more power and freedom we have, the more responsibility is required. We need the commitment to bring the same permanent characteristics to our buildings and towns today that stone gave to the great buildings of the Middle Ages.

We need to do it by designing and developing a comprehensive understanding of every area of our responsibilities. Not having to be replaced periodically reduces long-term material use and waste. This becomes a major factor in finding better harmony with nature.

At the same time, the design process must be free to discover and even invent new combinations and arrangements. It must capture those same meaningful qualities, special feelings, and sense of spatial design that create truly wonderful places for people to live and experience.

We need to build an infrastructure for towns and individual homes that is permanent, able to last for hundreds and even thousands of years. We need to design them with maximum livability and the flexibility to be refined and improved over time.

There's an opportunity here, and it will require entirely new concepts for our homes and our towns.

# Two Pivotal Design Issues
*Breakthroughs to make density desirable*

There are a few design influences that will fundamentally affect the design of any concept for a future town. It is necessary to come to terms with these before dealing with the needs of people and all that they do.

First we need to fully embrace the comprehensive consistently beautiful and functional qualities found in nature. This will be repeated many times — it's important. We tend to take nature too much for granted. This is a bigger issue than it may appear.

It is far beyond our ability to even begin to fully comprehend. If we pause with a deep, relaxed breath, just for an extended moment, to really appreciate the depth of meaning evidenced in some form of natural beauty, our most revealing and satisfying reaction is that we smile, … a warm, smooth, internal smile that flows over and through our entire body.

In some cities there are probably people who have never experienced that feeling. Being encompassed by the totally man-made environments of our modern cities has a negative effect on people. In many cities, nature is noticed most in the form of rust and deterioration, still the enemy rather than a friend. This is not some failure of nature.

In many cities, the only natural elements that still can be found are people, so it's no surprise people want to watch other people. There is beauty simply in how they move; even when they're bundled up to keep warm, they generate feelings of sympathy when huddled in some corner. Even if nothing is exposed but an eye, it communicates. So does all the rest of nature.

It's therefore no surprise that in some city conditions the only contact with nature is the interaction with other people. Perhaps it's natural that in whatever form nature is found, it is pushed, poked, and tested. Anger, hate, love, and power are expressed in all imaginable forms. Mind and crime games are played from the slums to the most sophisticated levels of society. Many people become as unnatural as the cities they live in.

But there's a basic problem: humans need to build large, town-size habitats; naturally, they will be man-made. The internal nature of our cities will always be automatically oriented to the needs of people.

Therefore, it's about a set of responsibilities. This is the first necessary fundamental commitment: Our towns must be more livable and provide more opportunities for people to understand nature and each other. The surrounding natural countryside should be in view from each home, as well as the main people-areas, and it should be close enough for daily interaction.

The second issue is cars. The purpose of the town is to provide for the needs of people, not cars. We are brainwashed into accepting the idea cars have a right to be in places they don't need to be. Some town layouts make them a necessity.

The needs of cars are different from those of people. Cars can easily be stored in private garages on a lower level. They can be as convenient as a few steps from each kitchen and a one-minute elevator ride. Even in this new ideal pedestrian town, cars will still have a purpose. They can provide convenient access to the more distant countryside, connect to typical roads, and be used in other, more-traditional car-based towns.

However, maximum livability can be achieved *only* if the town's circulation is pedestrian — no cars — and everything is within a short walk.

For some, accepting this as a design requirement will require one of those life-changing decisions. Previously conceived design solutions may interfere with some people being free to consider such a totally new concept. Even those who have written bad things about cars may resist the idea that a town doesn't need any cars at all.

So this is the second necessary fundamental commitment: No cars shall be seen in places for people.

Accepting these two issues as principles for designing towns will be a major breakthrough factor in making density desirable.

# Back to Basics

*Aware, but free from imitating the past*

The thoughts in the text so far touch on some of the issues that begin to encourage different ways of thinking. Being totally aware of but free from imitating the past is a technique for opening our minds to freely seek a totally new town concept.

Briefly: We need to invent arrangements and formats for a town that provide optimal opportunity and flexibility for every individual, home, and neighborhood; designed to look and function as an integral part of nature; optimum for every individual inhabitant in our time; and durable enough for hundreds of years in the future.

Part of being optimum is a having a reasonable cost. That means developing a concept where the individual home cost is about the same as a quality suburban home with comparable amenities.

Beyond that, the physical design starts on the equivalent of blank paper with the assumption of comprehensive responsibility and complete design freedom: no artificial limitations, no outside influence, no preconceived thinking, conclusions, or objectives, no popular past or current architectural movements or style, no cars or even horses as an influence on the town plan, and as if no previous concepts for houses or towns ever existed, except as further qualified by the following thoughts.

Whatever has been recognized as good — the best qualities of everything from past experience — is retained as a desirable objective, reference, principle, and guidepost. Not imitated — only the feeling, character, and quality rather than the specific details or spaces used to achieve them.

Every material, every color, and the understanding of space, however, along with how to apply them for certain effects, remain as part of the design palette. All of nature becomes a source for new inspiration and techniques to be borrowed, as well as the comparative measure of success.

For everyone to work together, seeking this new concept town, everyone needs to start from the same place. That means back to basics, starting from scratch.

*Part Two*

# Observations

*Part Two-A*
# General Considerations

# Where Did This Start?

*Most commercial intersections: ugliness*

This did not start out as an effort to design an entire town. It was born of frustration with the lack of design quality and livability in our typical towns. In the fifties, I would encourage friends to stand on any commercial intersection, look around 360 degrees, and simply notice the ugliness. Contrast that with a super-idealistic dream, that our challenge should be to build human habitats that did not imitate but rather captured the quality of being in natural places like the Sierras, ... the inspiration of elements like vertical tree trunks, draping branches, blue sky, tall outcroppings of rocks, flowers, and perhaps a lake.

There was the frustration of how little or how slowly a single truly high-quality architecturally designed building contributed to a community's progress toward a better overall design character, and how the responsibility for authoring architecture, good or bad, never received the same by-line credit in the local press as authoring a book or a photograph. This resulted in less knowledge on the part of the public and decision-makers.

These issues aren't as minor as they seem, but real frustration grew from questions like these. Is it even possible to design a city that could solve the problems and challenges put before us in 1961 by Jane Jacobs in *The Death and Life of the Great American Cities*? Is it possible to satisfy all the issues of affordability and transportation, as well as saving farmland and the environment? Is it possible to do all this efficiently and economically, and in such a manner that it would last as long as historically significant buildings from the past, permanent enough to justify the effort and able to offer the kind of livability where people would enjoy adjusting and refining it over time to meet changing needs?

Who would ever be in a position to take on such a challenge? Somehow, someone should at least try. Thus a seed was planted.

After 25 years, having designed all types of buildings with many design awards, international publication, and techniques for controlling building costs, I felt compelled to at least give it a try. I gradually phased out of normal architectural practice to struggle with the increasing challenges facing our cities, and that began the quest for ways to make my idealistic dream practical.

Over the years there have been improvements in residential developments, such as people-friendlier spaces, mixed-use streets, and better landscaping. But as others have written, the problems with human habitats go far beyond simply meeting the obvious needs of people and making places look nice.

With this totally new challenge, between occasional projects in the early research years, I studied, traveled, read, and tried to better understand history, nature, people, and what made places great. I realized great places didn't just happen and often weren't used as originally designed, but that individual users redesigned and refined them over hundreds of years.

Different countries gradually developed different paradigms in thinking appropriate to their place and time — differences like the hill towns in Italy as compared with scattered farmhouses on five acres in North America, or the concept of healthy fresh air that spawned the country house suburbs in 1800 England in contrast with flats occupied by extended families in Paris. There was no one universal answer.

I had a gradual realization that a town is simply one big building, spread out flat on the land. That's true whether residential buildings are one, four, or six stories. Streets can consume up to 40 or 50 percent of that flat structured space.

If we ever want to make efficient use of land, energy, and technology and to provide for easy improvement in the future, a better overall approach is needed. The only way may be to think of the design of a town as a comprehensive architectural problem, more like a building rather than just a planning layout on the land.

Only you and everyone known by you, working together, can ever hope to build a city that will solve all the challenges facing us. These issues will not go away — we simply have to find real-world solutions. There is a need for everyone to fulfill his or her unique role in designing and building this new concept town. So perhaps your thinking can join mine and eventually carry forward the thought process my effort on these pages attempts to lay out.

Before jumping into the actual process of designing a town, it will be helpful to review the factors on the following pages. These will influence the approach, refine it, and expand our understanding of the challenge.

# What Can Cars Tell Us?

*Give buyers a better choice: stop sprawl!*

If you want to stop sprawl, give buyers a better choice!

It may be interesting to remember cars before imports were so popular. There were a few automobile enthusiasts who recognized the waste and foolishness of America's giant tail-lighted monster cars. They preferred the engineering and fuel efficiency of foreign cars like the Porsche and its little brother, the VW Beetle; later came Toyota. But our manufacturers and buyers seemed happy. Imagine, in the early 1960s our cities were still filled with rows of parked cars in which not a single car was from a foreign country. As congestion, parking problems, and fuel prices increased, it took foreign auto makers to recognize the opportunity. They now command a large percentage of the market.

Can housing and towns learn something from this simplified example? Current housing and town problems offer a great opportunity.

Governments would never have been able to coerce or force Americans out of their car-buying patterns as effectively as alternative car design choices did. (SUVs are the public's way of dealing with a different problem: how frightened they are to drive. It's another argument in favor of a change, but how or what?)

Laws and ordinances, even when people are convinced there is a problem, will never get the general public out of their current home-buying patterns as effectively as a better choice.

The government is not the entire problem, and it's not likely to ever find a comprehensive solution. Regulations are based on what's been done before, refining what's current, and perpetuating the status quo. Even in a free economy, any new major breakthrough will face an uphill challenge. It's not realistic to expect the public to imagine and then vote for something they've never seen.

Any change also goes up against an entire industry, unions, and legal bureaucracy. If you expect to play the same game, any new idea needs to simultaneously satisfy all the players. But that will first require a willingness for all of them each to take the time to even think about it, … to seriously consider it.

The only way entirely new major projects of change happen is when someone is willing take up the vision and with a daring free enterprise spirit do it, like Walt Disney and his fantasy land, for example.

The home-buying public and the entire home-building industry are not going to change direction without incentives. New (Traditional) Urbanism is a step in the right direction, and it is the solution for many situations, but it does not attempt to address all the issues. It makes important incremental improvements to existing elements and it has basic incentives: its concept and decorative aesthetics are attractive to the market; it's attractive to the building industry, allowing them to save on construction and land cost. Those who benefit from New (Traditional) Urbanism have given it remarkable momentum. Their success should give encouragement to any other new concepts designed to solve more of our challenges.

Developing a choice that simply attracts the market may be a fundamental requirement, but even if it has more livability, that is not enough. There are other, more serious issues. A concept comprehensive enough to solve all the environmental and habitat issues facing current and future societies is needed.

Many in the market or business don't even recognize all the issues, much less care about them. Solutions to those issues have to be accepted as a responsibility by the developer of the end product — a developer with a greater purpose. Some would say hoping to find such a developer is simply expecting too much, that it's not likely to happen.

It will happen only if any new concept costs less and offers more. A greater vision needs more potential for livability and profit. It needs to offer enough obvious advantages that a large segment of the market will immediately recognize its value above all other choices

This could be an opportunity for a unique and very profitable untapped market. The conventional home-building market is good; like the auto makers of the 1950s, builders and buyers seem happy.

Perhaps, like the auto industry, the developer of the first prototype of any dramatically new concept will have to be someone who is foreign to the current home-building industry.

To address all the issues and make efficient use of the most appropriate technologies, an entirely fresh approach to how we arrange and build our habitats will be necessary. There is no way to get there by making incremental improvements to anything we are currently building. Examples of wonderful historic people-spaces already exist all over the world. We can learn from those. The design spirit of such places can be brought together and combined in ways never before possible.

The design concept proposed in this book attempts to present a comprehensive step toward an answer that is livable, affordable, sustainable, and safer, and in every way offering a better choice.

# The Horse Did It
*Concepts from before horses*

Throughout history, walking as the primary form of travel influenced the layout of early villages and towns all over the world. There are magnificent examples from the Middle East, Far East, Anasazi, Middle Ages, in North Africa and the hill towns of Italy or Spain. Streets and narrow passageways were primarily for people. Walking could get you to the places you needed to go. It was an experience filled with vitality and excitement. The street was a big part of people's lives and you were in the middle of it, not just a passing observer.

During earlier times in North America there was unlimited land, and you could claim part of it as your own. But you had to have a horse to get to it. And it was fun. Towns were designed for riding, hitching, and watering horses. As you passed down dirt or muddy streets, there was a subtle quality of separation, a sense of being protected by this big beautiful animal. Big streets became important. Distances grew. Cars came and did everything faster and better than horses. Streets were built everywhere. In some towns streets and front yards use over 50% of the land.

These are terms many exceptional writers and speakers have used to describe basic problems of our cities and particularly suburbia: sprawl, suburbia, cars, traffic jams, parking lots, freeways, streets, roads, alleys,

garages, strip malls, "uproar of signs," drive-ins, "stupefying ugliness," billboards, congestion, energy waste, smog, isolation, zoning, "look-alike subdivisions," destroyed farmland, lost time, "blandness," "no sense of space," "loss of community," "cartoon architecture," "grace-less," "centerless," etc. Cars are clearly a symptom, simply telling us that the layouts of our towns require cars.

Judging by the terms above, if this is the competition that a new town concept has to improve on, it doesn't sound like it should be very difficult.

So what's the problem? Why aren't there new concepts for towns and housing with great adventurous new ideas built everywhere? Most of these issues have been talked about for 40 years or more. Immediately after WW II, 400 companies were started offering better ways than using 2x4s to build houses. All but two went broke. We've also had government sponsored high-rise housing projects, but even those were generally social and housing disasters.

Making significant change to any status quo is tough.

Our process for providing housing is a giant industry; each component interlocks with the next. On its own, the process is not going to change in any significant way, and perhaps it does not need to. Even if it is challenged with an entirely new prototype, the market is so large the same existing patterns will continue for generations. Any new prototype will grow up like a new grass in a meadow. If it fits the condition well enough, it will be nourished and spread.

The most important feature of a successful new prototype is that it must be primarily for people, that is, it should be based on principles found in town layouts before streets had to accommodate the needs of horses.

# Being More Frank and Earnest

*Recognizing our situation*

We need to be frank about our situation and earnest in seeking understanding. We see so much variety in nature, yet each and every thing retains a certain commonality. Whatever the cause, ultimately becoming one with that commonality may be a basic factor necessary for sustainability.

Everything in nature appears to have near-total freedom of design, but with some amazing kind of instinct so everything that occurs results in harmony. All natural forms other than us, from crystals to flowers, know how to do that, but why and how we haven't begun to figure out.

No matter in what situation we find ourselves, facing anything approaching total freedom, whether it's environmental, social, political or economic, … it's easy for us to mess up. With few natural instincts, we have to figure out everything for ourselves. Families and cultures grow when they can pass on what they learn; there's too much for each generation to acquire.

When we fail, we try to blame it on the system, the one that gave us the freedom to mess up. It simply means we don't yet have adequate knowledge, understanding, and wisdom to match the freedom given to us. The functional and aesthetic mess-ups and shortcomings many see in our suburbs and cities is simply our society's report card on how comprehensively we fail to understand the necessary skills.

*Another objective for an ideal community may be a physical, functional, and social structure that is easy to comprehend. People could then under-stand how it all functions together through experience, starting at a young age. This could be a basic element required for a community's sustainability.*

Currently the scale and isolation of various elements of our cities pre-vent understanding. With the absence of adequate education, perhaps government's biggest disaster, our decaying cities have little hope of reinventing themselves.

Perhaps failure is nothing to be ashamed of; certainly shame will not give us the answer. Failures simply need to be overcome. Individuals and civilizations have been trying to get a grip on such things from the beginning of time. As most cultures matured, they came up with such things as the ancient list of the seven deadly sins, commandments, and other similar guidelines.

Today, few young people could name the rules of such a list, not to mention know how to apply them. Such guidelines apply to our relationships with nature, as well as each other. In recent decades, our culture seems to be making an almost conscious effort to unlearn such things.

It's our responsibility to know how to get along with each other and our magnificent natural habitat. In the long run, the nature of everything, all that is, … that's what's ultimately in control. It's our job to figure out how we can become an integral part of it.

As Frank or Ernest of comic strip fame once said to his psychiatrist, "It's not my grip on reality that I'm worried about —it's reality's grip on *me*!"

# Exploring Our Responsibility
*Unique freedom brings incredible responsibility*

Humans, by nature, can be adapting, forgiving, accepting, and satisfied (at least temporarily) in many different ways. This allows us to live on most parts of the Earth and make our own understanding of the beauty in it. Human tastes are malleable, shaped by our experience and associations. In very harsh circumstances the mind can relate to any beauty that is to be found. If able to focus, it can build rewarding feelings upon just a few small elements. Perhaps this human trait is the basis of our ability to survive, even to find enjoyment.

But within the depths of that same human makeup, there's a natural inner motivation; it's like being compelled to imagine improvements and finding great satisfaction if any are accomplished. Humanity realizes progress only if we act upon such imagined images.

In the design, structure, and function of all things found in nature there is perfection and harmony. In this regard, considering all known elements from the smallest to the most distant galaxies, here in our tiny space of time and on our Earthly speck, we are the only entity that makes things that are inconsistent with nature's perfection.

We have a special kind of freedom which brings with it an incredible responsibility, which we have never fully appreciated nor accepted. Why we have this freedom, we really don't know. Also, we uniquely have the capability for documenting our mistakes and our successes.

Getting even small things right isn't easy. Perhaps we need to be kinder to each other about the mess we have made of our cities and suburbs. Most of those directly involved have been continually looking for better ideas and choices.

Perhaps it would be better to think of all efforts as part of our training to eventually get it right, … a learning process. At least that implies eagerly searching, more openness to new ideas, and the potential for improvement—it gives hope.

We wait (patience is not the entirely appropriate word) during periods of stumbling forward and back in town design, hopefully inching ahead. It's similar to a child stumbling in the process of learning to walk, … not getting upset, just getting up and going on, … working together.

While we wait we are like captives, in many ways like slaves, to our current conditions, man-made surroundings, and the limits of our world of words. We're also trapped by our own minds—by limits on what we think and what we're willing to imagine.

Achieving the same level of quality found in nature is a monumental challenge. Perhaps it's the purpose of all the extra, unused, brain capacity science tells us we have. Can we somehow become connected to the same principles leading toward nature's level of quality? It can't be impossible. Other, supposedly less-intelligent life forms do it.

Maybe that's a fundamental responsibility if we ever hope to comprehensively address all the people, construction, and environmental issues.

# Thinking About Materials

*Domination over materials, no longer natural*

It requires the skill of an artist to build with natural stone: an eye for picking the right stone, perhaps visualizing several stones ahead as they are to be positioned with just the right effect. A stone wall can almost disappear into natural surroundings.

Then consider brick. Building that first brick wall must have been exhilarating. More people could do it, faster. It evolved its own special art form and skills: the graceful handling of the trowel, the brick, and the right amount of mortar shaped perfectly before the next brick was slapped into position. Eventually everything could be built in brick: buildings, streets, and sidewalks, like on the facing page. An entirely new environmental setting was invented. And with trees in planters and streams of sunlight, it could create a very beautiful environment.

Eventually many new materials were made. Entire cities were built, developing a new man-made aesthetic, one that didn't need to fit into any natural surroundings. Whole cultures grew, generations of people never seeing a tree in a natural setting. Magnificent and gigantic forms and spaces were created. Many people think of all this as being normal, even natural, a human expression of nature. But it is just like the brick, an art form all its own.

These man-made places developed as the seats of power. They controlled everything. They touched every part of human existence, from early education to music, magazines, and movies, a totally man-made

reality; often, completely out of touch with natural reality. So fully accepted are such norms that European cold-weather structures aren't considered out of place when they are built on a beach in Hawaii.

When we gained domination over our building materials, there was nothing to remind us that we were part of the surrounding natural world. If we had assumed that responsibility from the beginning, we might have done better, but it is never too late.

When we dominate anything we're in control. To be responsible, we need to know how to use it correctly, what's going to allow it to become the most it can be, to take advantage of all its properties. We need to know everything about it, what will deteriorate it, what it can resist, how it combines with others. Nature uses everything. Everything has a certain special value — its optimum application — and it is our responsibility to find that out. All of that is needed in designing for materials and opportunities for people to live.

Flying at night over cities sparkling with lights, they appear beautiful. But when the sun comes up, most cities look like giant bacterial infections overtaking the natural landscape. Some natural settings are blessed with tall trees that soften the impact of spaced-out, man-made boxes. But such spacing tends to consume too much land. Even some of those places, when trees drop their leaves in winter, look a little bit like a dog with mange.

At some point our habitats' damage to the environment becomes so noticeable it can't be ignored. Then environmental movements are organized. But here also, we have to be careful that these movements are not equally limited as captives of similar man-made thinking: over-organized, abstracted, and trapped.

It's easier to motivate movements and get money to treat obvious symptoms than fundamental causes. This means nothing ever really gets solved.

It may be unreasonable to expect us to discover any truly new vision by simply reworking existing processes that have been on a less comprehensive track for so long. We also need to be careful that General Plans based on our past do not limit inspired invention and new combinations.

Perhaps we need to think differently and on a larger scale. The image of natural rock formations on the facing page shows what appears to be a human-scale rock wall, but the rocks are actually the size of houses and those are full-size pine trees growing on the shelf. If the rockes are visualized as buildings, it certainly reminds us of the potential use of trees in cities.

This page shows rock walls in Japan. It is interesting to note how similar they are to the rock work in Machu Picchu on the other side of the Pacific Ocean. This underlines the universal nature of a material has and the power it has to influence how different people use it. The nature of their work is connected by the nature of a material.

From the beginning of our history, this could be the first time we have what's needed in order to accept the monumental challenge of integrating our towns with nature: livability with sustainability. Finally, we may have the necessary technology and experience.

If we make maximum comprehensive use of the best combinations of everything now available to us, it would be like having an entirely new construction material. This could offer new potential and solutions more universally satisfying than anything previously proposed or imagined. Within a hundred years, conditions may demand it.

# Our Future ... Faking the Past?

*Every other field has frontier-breaking innovation!*

Today, in almost every field, we have true frontier-breaking innovation. Much of architecture and planning, however, is in retro mode, trying to imitate the past or copying certain details rather than understanding the design principles.

We identify with the great spaces created in earlier periods, including the use of materials and shapes of openings. For any new town concept, people should have the freedom to build their homes with the kinds of things they have grown to appreciate. They should also have the freedom to easily change features as taste and needs change.

We have learned to do a few things well, like imitative surfaces and visual effects, but it's impossible for us to capture the entire past. Some of those massive stone walls have the patina of a thousand years. They have inherently sustainable characteristics.

What those early days didn't have is the technology we may depend upon for our long-term survival. This includes the way we arrange the major elements of entire towns.

We need to do our own unique combinations as effectively as they did theirs, with the limited technology they had. Stone automatically led them to build near-permanent buildings. For our time, if we develop durable designs, someday we might even be recognized

as good ancestors. Our contribution to the world might also last long enough to develop a patina.

No matter how effectively we can get the exterior skins of buildings to look hundreds of years old, or build exotic new sculptured trophy architectural edifices, neither of these will solve the challenges we face. Of course, there will always be a place for fantasy and pure decoration, carefully done.

Frank Lloyd Wright asked for honesty of materials. Our culture is still wondering how best to use them or what advanced modern materials should look like. Adequate contact with nature may provide insights or the needed balance in our lives.

Some think the stature of buildings that represented quality or power in the past can simply be borrowed through imitation, but past buildings were for a different drama and a different time.

Many of our popular buildings seem to be influenced by a Hollywood stage-set mentality and amusement value. Perhaps that's consistent with shallowness in some current thinking, where few appreciate anything as being real or true.

A major newspaper has a section called "Life," but it's mostly about an imitation reality — movie stars or amusements. The word "amuse" means "without thinking," like the word "amoral" means "without morals." Thinking is a fundamental reality of being human; does time spent being amused or not thinking make us less human than we can be, less real? Being amused can encourage either real or twisted thinking, re-creation or behavior patterning.

Is fake reality a symbol revealing a truth of our time? On the entrance to the University of Southern California film school a sign reads, "Reality ends here." We could put that sign on the entrance to some of our buildings and a few of our cities.

The separation from the reality of nature has existed so long in some big cities that people think it's the nature of human life on earth. They live like aliens on this planet. Some people live most of their lives without ever seeing or having a chance to enjoy a fully natural environment; to do that would require taking a trip.

Some tend to think human habitats and nature are meant to be separate spheres of reality. The paradox is many of those most concerned and wanting to be helpful in efforts to preserve natural places are those who are least familiar or closely associated with them. They often have little understanding of nature. In itself, perhaps that's not a problem. However, they also seem the least capable of understanding or being helpful in developing techniques necessary to interface adequately with it. Interfacing with nature may very well be among our biggest challenges for attaining a truly sustainable future.

The impact of our habitats on nature can be very damaging. This means we need to find the best ways to reduce the impact of every way we interact with things like farming, energy, materials, water, sewage, recycling, reuse, and continuing use. Everything we have, make, or do is taken, borrowed or copied from nature. We need to seek concepts that are about putting back as much as we take.

People have not changed that much over thousands of years. During that time they have built and maintained the places that have served them best and that we have grown to appreciate. It's not necessary to abandon that.

But there's more to it than faking appearances of things from the past. To have a realistically sustainable future, new concepts for towns will require true innovation.

# Environmentalists vs. Developers

*No current solution meets everyone's objectives*

Fights exist because, until now, there has not been a proposed solution that had the potential to meet everyone's objectives.

In many communities, disagreements are not about whether housing is needed, they are usually about how and where to build it. There are valid concerns on each side. For example, developers want to meet the market needs with reasonable construction cost and pricing, while environmentalists are concerned about pollution and using up good farmland.

There's little to be gained by having battles between various housing development and environmental groups. Sadly, with some it has degenerated to a form of visceral hatred, implied in what each side says and writes about the other. A special vitriol is added when the other side can't hear what's said.

Once that happens, people join opposite sides, become entrenched, and cling to what they know or have been told, certainly not open to any new thinking. Often they have just enough information to stir emotions.

Reminiscent of the Hatfields and the McCoys, people often seem to wear blinders allowing no side vision. They view everything through a screening device, permitting only things that fit a preconceived form to pass. No truly new ideas get past that screening. Stylized catch-phrases with emotional attachments become tools for influencing thinking and mobilizing a group effort. Some don't really understanding the total significance of what they repeat.

Compromises, when achieved, rarely provide a totally satisfactory answer. Rather than exchanges becoming a creative process inventing comprehensive solutions, they tend to be patched-together piece-meal elements of each special interest and things easily visualized from the past.

To make matters worse, things can be muddied by hidden political or economic agendas. For some, planning is simply about creating or strictly following words found in policies and ordinances. Many have no training for visualizing the effects or interpreting those words in ways that have any connection to practical or physical reality.

45

Occasionally there is a glimmer of hope as various sides come together, realizing that all have something to offer. At best, the challenge of evolving or inventing a new town concept is probably too complex to be solved with any normal process.

New parameters must be set. Often the correct questions or issues are not even raised. Much longer-term objectives must become the foundation of every discussion.

*Inventing a totally new comprehensive approach to human habitats designed as an integral component of this earth is as complex as sending a spaceship to the moon.* Similar systems are involved, but addressing all the human needs actually makes this much more difficult.

However, being invented is only the first step. The second, and perhaps the most difficult, will be getting all those who have been in the battle trenches for so many years to open up their thinking, take off the blinders, and abandon their screening devices, ... to be free of preconceived thinking.

Then, if they are willing to allow their minds to explore possible combinations they never before imagined as being appropriate and without drawing any conclusions until the entire potential is comprehended, ... only then may they be ready to consider solutions adequate for our future. Many of the challenges we face are the same as before but many are new and very different.

That second step is necessary for anyone to have the freedom to imagine wandering through the new spatial environments described in the rest of this book.

This is about an exploration, beginning within the mind of each individual, to construct a new environment. It's a process of discovery into entirely new territory, a new frontier. Think of it as similar to Lewis and Clark opening the West. It's like discovering a new kind of land, ... places to build with new potential. Like being there, not just as an observer, but as a contributor within a format were everyone has the freedom to build the home of their dreams.

Hopefully this will become a shared quest, with each reader seeking ways to add their own newer concepts or refinements. As each visualizes wandering though these proposed arrangements of spaces, comprehension and spatial understanding will grow; they can note the places where their favorite interest can be included.

Everyone's special expertise from various backgrounds is important and necessary for success to be comprehensive, … a concept with the potential to meet everyone's objectives.

The best solution will likely be one that provides a framework where a maximum amount of freedom for each individual home is possible.

*Part Two - B*
# Circulation and Change

# History & Streets

*Learn from history, but don't be forced to repeat it*

History suggests there are only two approaches to the layout of towns: grid streets and random streets. We are supposed to learn from history, but we are not limited to repeating it.

Our goal is to capture the essence, and the optimal features for livable streets, first seeking the highest possible design objectives for the individual, each home, and each neighborhood, being open to entirely fresh concepts; let it evolve. If any of those objectives finds conflict with cars, and certainly some will, put the cars somewhere else. The entire town should be designed primarily for people, walking. If you think about it, for a modern town, this probably has never actually been possible before now.

In two-dimensional town layouts, the only way to move to every location is with some kind of street pattern. For people it could be narrow and irregular; for cars it's bigger and straighter.

For people, we now have the opportunity to build an entire town in a three-dimensional framework. Elevators could go to all levels. A compact town layout would make it more economical to connect every home with every imaginable wire, pipe, cable or tube,… eliminating a lot of transportation.

It would be simple to have a different vehicle circulation and parking level below the people level. That's common in many shopping malls. Parking and service can be located under the entire town in relation to what is above it.

With streets just for people, we should try to understand the best elements of traditional streets.

There are many streets in the world that have become famous. They have been the setting for stories and movies. Some have great historical significance. Some have been thought of as great streets.

There are books about streets. *Streets and the Shaping of Towns and Cities* by Southworth & Ben-Joseph is an excellent, well-illustrated reference. *Great Streets* by Allan B. Jacobs is a wonderful tool for helping to understand some of the characteristics and features that seem to contribute to great streets.

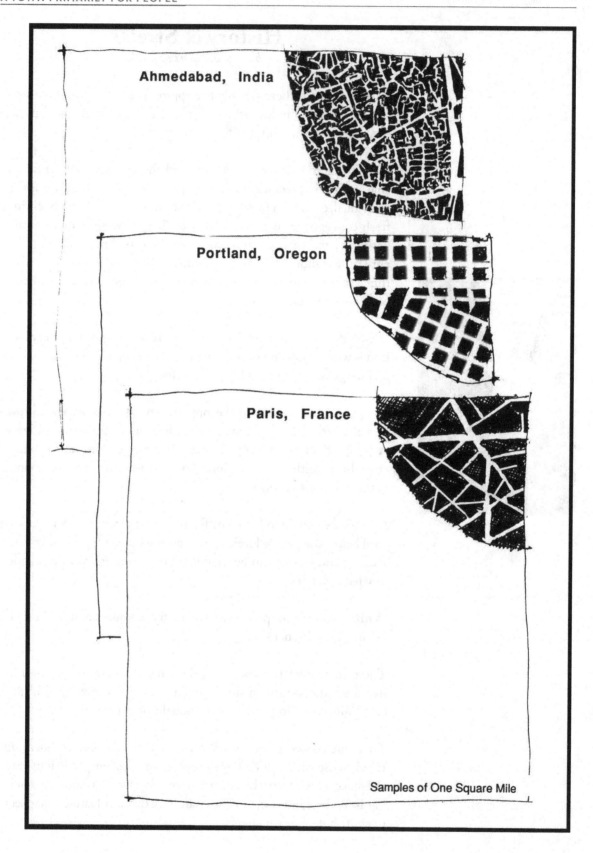

Ahmedabad, India

Portland, Oregon

Paris, France

Samples of One Square Mile

The latter book has excellent sketches and maps showing the graphic patterns of many towns and their streets. Each scale map shows a square mile, which makes them great for comparison. The new concept proposed in this book also shows a square mile layout. These are helpful for comparing land use with a typical square mile of suburbia.

The oldest towns have totally random, small-scale curving pedestrian patterns with no obvious preexisting site influences. Like Ahmedabad, India; New Delhi, India; obviously just for people. Right angles do have a subtle influence.

Venice, with a river randomly curving through its center, has straight street segments trying to create 90-degree intersections, but all angles are adjusted to adapt to the curves of the river. Rectangular stone or brick construction must have been a factor.

Look at Cairo, Egypt; Lucca, Italy; old Barcelona, Spain; Rome, Italy; Seoul, Korea; Zurich, Switzerland; and Tokyo, Japan. Each has its own unique subtleties and a human scale but a little more regular pattern.

The newer parts of Barcelona, Spain; Portland, Oregon; and New York, New York show the boldness of regular grid patterns and how much land is used.

Paris is uniquely different. It appears they realized straight streets were best for wagons and diagonals are more efficient to get directly to where you want to be. Personal decisions must have laid out the roads rather than some arbitrary preconceived planning concept or grid pattern. Some say it was intentionally designed to confuse the Romans.

# Street Participation

*People make it art ... Great!*

The only way to appreciate a street is to experience it. There are a few features we can identify that enhance the experience. The shape of the spaces and design elements give character through details, materials, quality, light, colors, plants, and, of course, people. But it seems you can have all those and still not have a great street. Or take most of those away and still have a great street.

It could start out as the worst space you can imagine. It could have very tall barren plaster walls with only one other opening off to the side, few overlooking windows. Add a small ray of sun falling on colorful umbrellas over a few tables with chairs, a few people seated, some walking by ... perhaps fresh tempting smells, muffled sounds of a guitar from inside. It may not always be at its best, but when it is, people will come and enjoy it. When everything is just right it becomes a great street.

A great street is a work of art. Everything works in concert. There seem to be no absolute rules, no guides to follow. But there are intuitive rules, and over time even ordinary people discover them. In those moments, they become artists. All this probably sounds very abstract, beyond our understanding, but it's real and possible.

The one element that adds to making almost any street great is people, ... simply by being there and doing what they enjoy they change

the street throughout its history. The vitality of the town or neighborhood will be reflected by what they add, change, and refine. Any new town concept should allow people to finish it over time.

It can be a narrow street, medium or wide. A wide street may need lots of people, but the same street, with angled dim first morning light, a gentle rain, the wet surface of interesting paving stones, and at the other end only one person holding a bright colored umbrella moving through the space, turning, looking, pausing, … at that certain moment, even nearly empty it can also be a great street.

Great streets are a stage for people, a living stage of many vignettes of life unfolding.

For the participants, an active street seems to help carry over to the next day memories of excitement from the evening before, but each day is fresh and new, and once again its drama grows. Some things are the same, bakery smells and coffee, chirping birds, a door opening, a breeze causing little tinkling noises, then sounds of movement, water running, people greeting, gradually more activity, and before you know it the street scene is alive. A street should be expected to be a place for people to use and enjoy.

But if so many people are participants on a certain main section of all the towns' many streets, who's on all the other streets where all those

people live? What we find is many are lonely neglected streets most of the time.

All the above factors become reasons why a town's design should give everyone the opportunity to be a participant of Main Street as often as possible. Actually, it would be best if there were no other streets.

There are many ways to participate. Of course, walking to local destinations is great, the more often the better, as well as sitting at cafés or benches along the way. But wouldn't it be great if the front porch of every home overlooked Main Street? This would add a new dimension to how great streets could become.

# Grids and 2-D Plans

*Pedestrians walk in all directions*

A walkable town without cars can have pedestrian street patterns running in any direction.

There were grid street patterns before there were cars, and a few even before horses. There is evidence of grid patterns in early military encampments of Alexander the Great and the Romans. Efficient layouts of water pipes or property lines and building materials with rectangular shapes were later factors. In some cities, grids aided in defense. Napoleon intentionally cut across Paris's unconventional street patterns with straight radial streets for moving troops and cannon sight lines, diagonals like the Champs Elysee, now a great street.

Many towns in the United States started with a town square in the center and grid streets, but beyond that the patterns that developed around the towns were often the result of cow paths. This all changed with the invention of magnetic surveying and as towns grew in the West. After that, almost all town patterns were based on 90-degree grids, even where they were turned to be parallel/perpendicular to a random railroad angle.

In a town designed primarily for people and for encouraging them to walk, a grid street pattern does not seem to be a necessity. If you wanted to walk to the opposite corner of a mile-square town with streets in a grid pattern, you would have to walk two miles to get there.

Two-dimensional layouts for towns, even where they are not grids, have other inherent limitations. Current town-planning approaches may be fundamentally inadequate for future challenges. Three-dimensional arrangements of all the complex activities of a town could improve and increase opportunities for better functional and livable relationships.

Consider that a town is more complex than a tree. Imagine if we attempted to diagram and label all the complex biological functions of a tree on a flat piece of paper, but could only draw straight lines in grid patterns to connect all the boxes. It would need to show each overlapping

55

element and everything connecting them, with boxes labeled for leaves, connecting stems arranged for collecting sun light, boxed-in groups on branches shown as arterials, then the major freeway-sized trunk connected to the root system of more arterials branching off with different groups of boxes, etc., … on and on as a tree grows.

This is, essentially, what aerial views or land-use street maps of our towns look like. They are simple diagrams of activities, limited to 90-degree street patterns. The overall format of what we build as towns is still in the two-dimensional diagram stage. It may have been good enough for primitive encampments, but not for the many dimensions of modern life.

Diagrammatic 2-D towns could never reflect what a living human habitat should look like, any more than the 2-D tree diagram would look like a living, three-dimensional tree.

Once cars are no longer part of the pedestrian areas, entirely new designs and arrangement opportunities open up. Pedestrian streets and walkways are friendlier to the enjoyment of front porches as functional living extensions of the home. That openness can expand to include the walkway and neighbors as they pass by. This will even

be a big influence on how we design the individual home's interior, helping it to function in concert with the exterior.

More compact clusters of homes are possible if we do not have to allow room for cars moving in the same areas needed for children playing and people going places. With no cars the usability of space is suddenly enhanced. Then the design focus can be primarily on how much space people actually need or want.

Without cars, entirely new arrangements between homes, front porches, walkways, and play areas become possible. An entire neighborhood cluster could have its front overlook the main pedestrian shopping and entertainment street. This could be located on an elevated terrace, giving a dimension of privacy from the more public character of the street.

These are just a few possibilities. Three-dimensional town arrangements provide entirely new opportunities for how spaces can relate to each other and people can live.

Grid streets and even totally irregular pedestrian pathway patterns are patterns from the past. They were for a time when stone buildings had to be placed on solid ground. Current construction techniques allow new opportunities. Three-dimensional arrangements of activities in a town create patterns of circulation that cannot be fully shown in two-dimensional plans on a flat piece of paper.

# Boring Streets Have No Porches

*As found in many towns*

No one is going to sit on a front porch for any length of time if very little is happening on the street. Houses without porches did not happen by accident. There just isn't very much of interest going on in most of our neighborhoods.

In small rural towns the main street was probably the only way for travelers and farm-to-market traffic to get past that particular section of the country. That's where the world passed by and there were porches for watching it.

One of our problems today is that there are too many streets. You can find miles of streets in almost any city that are boring. If people walk at all, they walk to and from work, passing by no more than twice a day. Some streets in large cities, with even six stories of housing above commercial, are essentially empty of pedestrians most of the day and especially after hours. Then those commercial spaces tend to be leased by businesses less oriented to people. In some cases the sidewalk and street traffic use does not even justify the maintenance. Eventually such streets become dirty. They can become a liability rather than an asset.

As another example, it takes at least four to six blocks of four-story housing to financially support one block of shopping.

To support a shopping area with enough shops to meet people's needs, a town ends up needing lots of residential streets, often too spread-out for effective local transit. Many residents end up being located a significant distance from the shops, just a little too far to walk; then weather easily becomes a factor. Where cars are permitted people will be inclined to use them. The shopping blocks will have vitality only if people make the effort to go there. When business is good and traffic gets crowded, romantic shopping areas with quaint narrow streets become less convenient.

Once people are in their cars they are easily distracted to other shopping opportunities and they will use the less-busy residential streets to get there. The play-area potential of those streets is then reduced. If barriers are installed to direct or slow traffic, people get upset and the entire town suffers.

If cars are eliminated, it will be necessary to make the shops convenient and accessible from housing only by walking, but then all of the housing will have to be closer to the shops. More people walking will make the streets less boring.

In conclusion, an effective shopping street needs lots of pedestrian traffic. This will make for a more successful, livable, and interesting street. Ideally, the best place for all the supporting housing is with its front porches overlooking the shopping street with all its activity. People on front porches will also see each other, generating new dimensions of livability. This can become the foundation for an extended neighborhood and the community.

# Currently Popular Solution
*Still based on cars; nothing is quite close enough*

New (Traditional) Urbanism is one of the better refinements of community development for certain locations, if done to an adequate level of quality. It reduces the amount of land needed and developer costs. In contrast with an entirely pedestrian town, however, neighborhood circulation layouts are still based primarily on the needs of cars.

In existing towns where small infill spaces provide the only opportunity for housing, New (Traditional) Urbanism is a reasonable choice. When it's possible to assemble infill properties to be a hundred acres or more, however, the advantages of a three-dimensional town concept should be explored. It will likely provide better opportunities for handling people's current need for cars, as well as the transition to a truly pedestrian town in the future.

New (Traditional) Urbanism is still just a little too spread out on the land. It doesn't allow adequate potential for all the fully integrated, efficient, or comprehensive energy savings or waste and recycling systems. Any additions or changes still require digging up the street. The streets, being exposed to weathering, have the typical periodic maintenance and replacement cost. It will be necessary for any truly sustainable long-term town concept to solve these fundamental issues.

New (Traditional) Urbanism gets very close to solving many of our challenges. But to have enough housing for a viable shopping area some people will still have to walk a little too far. Things are almost,

60

but not quite, compact enough for walking or local transit. Where built, most people still rely on the car a great deal. Not everyone has close enough visual or physical proximity to intuitively feel part of an extended neighborhood or the community.

It can have a pleasant appearance, and by grouping three- to four-story houses closer together it can offer more open space than typical subdivisions. However, except for that, in terms of the overall impact on the land it is still basically what we've had before.

Unfortunately, a couple hundred years from now we will be facing the same problems. Our housing stock will be wearing out and too expensive to maintain, and viewed from above there will be housing as far as you can see. Usually the open spaces are only recreational, not part of any comprehensively integrated system. New (Traditional) Urbanism is a refinement over past designs in the U.S. and Europe. Once built out in an entire area, it may not be much better than the L.A. basin is now, just a different arrangement. Even with its open space, the housing is just a little closer together, a little denser, and a little taller. But it's not really tall enough to give great views or any particular advantage. At some point the concerns about encroaching on farmland will come again.

Most New (Traditional) Urbanism projects should be called New (Traditional) Suburbanism. Streets are not convenient enough nor is housing dense enough to be called urban.

New (Traditional) Urbanism's biggest advantage may be that it improves on car-oriented planning solutions, but it also helps to perpetuate our major problem. Perhaps it is the best a car town can be. It uses less land and makes construction costs cheaper because of shorter streets and sewer runs. This is all good in the short term, but it never really improves everything enough. It makes walking more attractive but not quite convenient enough to support a commercial area without cars.

It's a step in the right direction. It's needed. Currently it is the perfect solution for certain areas in many cities. Until now it's been the better choice.

But city and county general plans that *prescribe solutions rather than objectives* will prevent truly comprehensive solutions from being developed. Then, once again at some future time these cities will find themselves facing the same housing development trap.

New (Traditional) Urbanism solutions tend to be built to last a little longer than typical construction. They will be harder to remove and replace when something comes along that will actually solve all our challenges better.

# Total Change Steps
# On Everyone's Toes

*Every talent is needed for a new concept*

Many readers have probably already taken offense at critical comments made about their particular areas of expertise. But every area of experience brings with it special understanding, and each will be needed to assist in testing and refining any other new concept.

While it's obviously not the purpose, any truly new and comprehensive concept will end up changing some portion of everything involved in the current town and housing industry. Everyone's toes get stepped on at some point. Totally new thinking shouldn't be expected to fit any known pigeonholes. Unless we are willing and able to think freely, there can be no method for dealing with totally new sets of ideas.

Perhaps a good starting point in efforts to solve our challenges or develop any new comprehensive town concept is to consider the following comments:

- Environmentalists know things developers need to know if the desires of each are to have the fullest potential for success.
- Developers know things environmentalists need to know if the desires of each are to have the fullest potential for success.

The same can probably be said for any particular area of expertise relative to all the others. It's about overlapping experience, knowledge, and working together. A totally new town concept is a complex undertaking and will require every known expertise and a few new ones.

This is not about creating a utopian vision. It's simply a pragmatic process of looking at every need and every component, then seeking the best combination of optimum techniques or arrangements to satisfy every need. There's no reason or time for anyone to be offended.

Everyone can add his or her input and join in the excitement, the mystery of the search, and the quest for new concepts, inventions, inspiration, and new potentials yet to be discovered.

*Part Two-C*

# Connecting and Walking

# Nature's Connections

*For a town to be all it can be*

George Lucas was probably right when he indicated humans have always thought they understood the world around them; that on a numerical scale the cave man maybe understood at level 1, ... we are at about level 6. What no one fully understands is that the scale is a million.

It's good we have some inner self-assurance or a bit of overconfidence and short-sightedness as part of our nature. Without that, we would have given up a long time ago when we realized how much we don't know.

We've only scratched the surface of all that is. Chandra X-ray Observatory looked south recently into a tiny part of deep space that appeared as totally blank sky. When magnified, it had almost more stars and galaxies than could be counted. We are finding the same endlessness in the smallest things around us—always something new. Evidence now suggests the Great Pyramid was built in about 10500 BC rather than 5000 BC. Our biggest unknown is probably ourselves.

Some fear all this growing awareness. A few are beginning to seek comfortable thought patterns that allow them to throw in the towel, quit seeking, retreat within themselves, zone out, or take whatever limited quantity of knowledge they now understand and hide in some corner. This kind of thinking has caused some countries to stagnate for centuries.

Everything we know about, except us, functions as if connected to some incredible automatically orchestrated system of interaction and communication. It's all around us.

You would think one of our greatest efforts would be to develop an understanding of the source and nature of any possible connection with the world available to us. Instead, there seems to be conscious resistance to even making such an effort. We consciously set barriers that limit what is worthy of our exploration. Such positions are based on what we think we know, and there's continuingly growing evidence showing a greater vastness of what we don't understand than what we do.

You would hope that as different efforts at discovery made any progress, we would each share successful techniques at making connections and work together to refine our understanding. Instead, we ignore others' discoveries and fight to support our own, even if limited. Some even suggest zero effort in this area is the only reasonable action. We seem to forget that complete truth will be true no matter what any of us think. And it will endure, leaving us behind if we so choose.

We have a lot to learn by simply trying to understand and imitate what we observe in nature. *Biomimicry* by Janine Benyus may encourage something more comprehensive. Perhaps exploring and attempting to mimic connectedness in nature could help us better fulfill both our needs and nature's.

Could it be that inspiration is a moment of connection? Curiously, at perfect moments in history inspiration has given us entirely new inventions, and even though we may not entirely understand why or how they work, we can use them. Obviously, all things in nature that appear connected don't know everything about what they do, … but just enough to be all they can be.

Do we need that kind of inspiration to add the capability for us to make a town be all it can be? Does our freedom give us the potential and therefore the responsibility to attempt to be more, know more, and do more, … including connecting?

Could future generations of children living in a conceptually new town that demonstrates such connections have a better understanding of nature and each other? Could lacking this be part of our social problems today?

Digging a little deeper into the connectedness to all that is may seem like a subject unrelated to concepts for new towns, but it may be a fundamental element for attaining maximum livability.

Is it possible that our most advanced computers, being increasingly subject to smaller and smaller forces, could become more connected than we are? But they would never have the freedom of choice we have. The only choice we don't have is not to have a choice.

# Thinking From The Beginning

*Computers have two variables, nature starts with six*

We take everything so much for granted. If we were connected, as all of nature seems to be, we would probably take that for granted too.

On our Earth, we tend to believe the possible conditions for thinking exist only for humans. With all our known limitations, can anyone seriously believe that thinking happened only for an entity like us, for the first time — *ever*? Can that continue to seem reasonable, considering all the other incredible new things we're discovering about all that there is?

We are now beginning to see the grand possibilities of computers. Our old ones do all they do with only two variables. We are finding great potential advantages to computing with more variables.

From the beginning, nature's physical world started with things like quarks, the smallest known particles. Quarks have six variables. There are also simple examples of atoms, like hydrogen and oxygen, that remember how to act when they get together, and poof, there's water. Could that demonstrate a primitive kind of memory? From quarks to galaxies, we see evidence that could have resulted from memory, communication, interaction, and symbiotic relationships on scales we can't even begin to appreciate. Some we've only begun to notice. Are there other, more bizarre possibilities?

*It's impossible to observe the process of thinking.* How can we assume, starting with six variables at the beginning, that quarks could have evolved into all the other stuff without evolving systems that can think, or at least communicate. We can't assume that everything is limited by the limits of our thinking or understanding.

Could it be we are very much like an Aborigine surrounded by radio signals but without a radio? Or surrounded by stacks of CDs containing the entire history of everything humanity has ever known and no CD player? No receiver means no reception. The only thing we can be certain of is that we are not connected nearly as well as everything else in nature.

If we are using less than 15% of our brain's capability, everything we think is obviously more limited than our potential. And if it's not used, why did it bother to evolve, anyway? It's also interesting that brain monitoring, as reported in *Science News* or *Scientific American*, shows outward-seeking meditation (as opposed to inward meditation) apparently activates a larger portion of our brain cells. Is the rest of that extra brain capacity just waiting to be plugged in, connected?

Imagine: If we eventually develop a fully thinking robot, the only way we could allow it to be fully free and fully thinking would be to sever all possibility of any connection to us. We would have to do this to avoid any possibility that we would influence or contaminate that freedom. If we interfered with that freedom and the robot became clearly aware of us, it could be so overwhelmed that it would probably become dependent. The only way it could avoid being smothered would be to exist on its own terms and within its own understanding. The only way we would be sure we had developed a fully free robot with comprehensive intelligence would be when, through its own conscious effort, it became knowledgeable enough to actually discover us.

Imagine, from the beginning, if there has been an evolving thinking process, its influence could have been at the smallest electrical force level, possibly before the physical and thinking worlds developed in their separate directions. Perhaps similar to the role of a software program, it only needed to influence the variables. Could this suggest an explanation for the pragmatic yet playful creativity that seems to appear in nature, an evolutionary process that could be manipulated? Atoms and DNA can be manipulated—even we can do that.

This suggests an alternative that pushes the limits of or is contrary to what many of us believe about the nature of things. Most agree an evolutionary system of random chance arranged everything, but, similar to the auto/suburbia/housing industry, such systems are self-contained and self-reinforcing. Neither is free to generate alternatives to itself. Is this a potential flaw with any system?

Newton's theory of color worked fine within our then-dimension of understanding and need; no one realized other potentials were missing until we discovered there are other dimensions for perceiving and working with color. But that required open-ended thinking.

Since the beginning, humanity may be the first time the physical world and the thinking world are joined in one entity, like the most advanced robotic computer and possibly including a transceiver. Humanity appears to be the only life-form that replicates, thinks, learns, talks, remembers, and has freedom of choice, and is thus able to think and act directly in the physical world. Rather than acting alone, it may also have the potential to be connected to that thinking dimension, … to network with it. By virtue of this combination, the unseen thinking dimension would be able to extend its capabilities.

The world of that thinking dimension could envision things beyond what it could do alone, but by using humanity it could do more. Note the difference between how a bird can fly vs. a jet plane. In airplanes, even though originally somewhat primitive, humanity rearranges components from the physical world in ways different from what could grow in nature.

Is this potentially a new level of synergy for symbiotic relationships—the thinking world combined with humanity taking the physical world into new dimensions—a pattern for the future? Is that sort of what we're trying to do by combining inspiration from nature to help us achieve our dreams? We may need combinations of all dimensions to yield fully integrated human habitats.

Looking forward, we are already considering biological computers that can grow themselves. If our curiosity remains open enough, we may gain more insights, such as what elements would be required for

thinking to have been there from the beginning.

To seriously consider these Star Trek-type ideas requires a high degree of openness. It is understandable that for some in the current housing/town industry to seriously consider entirely new real-world housing/town alternatives may take a higher degree of openness.

# Town Centers, Age, & Flexibility

*How long should a town center be designed to last?*

For how long should a town and its center be designed to last? Has anyone ever asked this question before? Probably not, why would they?

Most towns exist in a location because of something that was there before it was a town. Often they are based on a major circulation route or joining of routes, like highways, rivers, or ports. Maybe it was a beautiful setting or a flat piece of anywhere (or nowhere) central to a group of settlers or activities.

Once established, it becomes a magnet to all sorts of town-related patterns. Everything over time added together increases its importance as a location.

Almost everyone knows the common events of gradual change over time. Buildings are built for needs as they occur. As towns grow, new needs grow more buildings. More boxes are added and stacked higher while others are removed. If the old town hall center has enough stature, sometimes it's kept. Then, as needs increase, the larger new center structures are often built to the side or somewhere else. If the original facilities are too small but in the right location, often they get torn down, so the old town center's original character is gone.

Surrounding buildings often become outdated and are not easily changed. They go down in value, and tenants move out to newer buildings. Eventually the town center is further weakened by automobile parking lots and driving distances. Sometimes it is a blessing if the old center is simply abandoned, with buildings left empty or with low rents. Then at least there could be something of significant historic value left to save in the future.

Variations of this have happened across the United States. This is not planning. Sometimes it is desirable and sometimes it's not. The reasons for the original town's location are often still there; the beauty has simply been hidden or destroyed by how and what was built.

How can such abandonment or loss be avoided? Every city or town is different. In many cases it will be impossible to recapture the past. Long-term thinking about the future, however, would give us more options for gradual change that can justify preserving historic structures and locations. In some cases it may be possible to once again surround them with open space or farmland, if that was their original setting.

All observations from our past will help in inventing new thinking. Below are a few very broad concepts to expand our thinking as we search for options. Some of these may be appropriate for certain existing towns.

Farming and open space can be an integral part of town/housing recycling systems. This will make it practical to bring open space into old downtown areas of existing cities with large infill sites.

Town centers should have a strong presence of place. Any new or restored town center must be built to be permanent. Then what others build around it can be built of the highest quality with confidence that the investment will be protected. If the building's framework and infrastructure is permanent, long-term bond-type financing with lower rates will be possible.

But how do we accommodate change? Simply design buildings that are flexible enough to adapt. Eduardo Catalano, FAIA, formerly head of the graduate architecture program at MIT, believed all major buildings should be large frameworks that allow anything to be built or changed within them.

Once the structures and roofs of durable systems are properly finished and everything is protected from the weather, they are essentially permanent. Interior spaces, two stories high between utility floors, allow for complete flexibility. They could be divided in appropriate sizes, allowing for any function.

None of the housing we are building today is affordable for the long term. At some point, all of it will have high maintenance or replacement costs. A permanent infrastructure may provide the only way we can ever hope to catch up with our housing demand.

Civic center buildings should all be constructed with the same framework methods. Then, if any need to expand, they could count on similar kinds of space being available nearby.

All the relationships between structures within the town center can be carefully designed, creating high-quality, permanent massing and spaces integrated with functional systems of open spaces. Being permanent justifies the extra or special design attention. Housing would be an integral part of the area, occupying similar frameworks.

The city center would be its own complete town district, including open space with its reprocessing systems. It must be complete to compete. The design should have enough concentrated density of housing to make transit economically practical.

Many generations of a friend's family occupied the same home in Spain for over 500 years. After the initial cost, think about all the savings. That's affordability. That's sustainability. That's historic preservation at its best.

If we can find a way to do something similar, that could be the biggest sustainable advance in affordable housing this country has ever made. Homeowner costs will be less than comparable current housing over time. Infrastructure that is protected is easily maintained. These are sound objectives.

Even in our current market, finding a way to build a town to last hundreds of years maybe our most practical alternative. Cost savings could justify that, but the lifestyle benefits could be even greater.

This is a brief outline of several interrelated subjects. The relationships will sound less abstract when properly interfaced with all the other elements of the town. Basically, the interior spaces of the entire city should have complete flexibility and the major portion of the cost should be the permanent infrastructure. Then far into the future, the on-going cost of using buildings and housing would be reduced by 50 percent or more — permanently.

# Can't Get There From Where?

*Current planning traps any significantly new idea*

Designing the layout of a town rarely starts with a completely blank piece of paper. There are development standards, zoning ordinances, and so forth to give some idea where to start. If not for all that data, every time we faced building anything in a town we would have to reinvent all the standards. They are based on tried and tested experience. They determine the basic nature of every town.

Problems stem from several of our fundamental concepts about towns. Most of what we struggle with and fight about are symptoms of a basic town concept. Typical issues are space, height, traffic, affordability, and farmland issues.

A new, comprehensive vision could anticipate and overcome those symptoms from the beginning.

"Car-fully" designed towns are an obvious symptom, a predetermined parameter. Town layouts meet the needs of cars, so cars are needed to meet the needs of towns.

The planning process has many similar circular relationships reinforcing each other, continuing around and around. They grew for centuries, are taken for granted, and are considered normal.

Perhaps the most fundamental contributor to the problem is our two-dimensional subdivision of land. It usually is the guide designed for towns spreading out over the countryside. Anything built can be seen by everyone, particularly where there are few trees. This creates symptoms we make efforts to control.

Typically the size of a house is controlled, which controls the type of neighborhood and who lives in it. Family needs change, rules prevent changes to the house, people move. That requires continual building of new houses, plus fees, commissions, moving expenses, and so on. This amounts to added and recurring costs to the home-owning family without any additional value.

To protect adjacent property values, you can't simply build only what you need or can afford. And you're not free to rent out what you don't need. Those seniors with homes now too big have to leave their cherished neighborhoods. People with low budgets are forced into older or lesser-quality houses and neighborhoods which have high maintenance costs.

When towns first started there weren't such controls. In the flat lands of the West, people bought a lot, maybe dug a hole for a house, and then over time built when or what they could afford. That's about as natural as you can get, and that's why we ended up with controls.

We must recognize that our current town-plan concepts do not adequately accommodate affordable housing; it usually needs to be artificially supported. Ways to avoid this can be part of a new town concept.

All those planning standards and their words are continually revised and modified, but they are still based on what has been done before. The words are many, and they continue to grow. Those schooled in growing and administering that process don't have any inclination, reason, or tools to change the process. This is also a symptom of the basic problem.

Our current thinking is trapped in a self-orchestrated whirlpool of bureaucratic complexity that drowns any significantly new idea for change before it surfaces. And that's because no one ever has a chance to address all the problems at their root.

This process extends from the individual carpenter through general plans, financing, and material manufacturing to the highest levels of government rules and politics. Home-building is very important, and perhaps our most extensive industry. It's not all bad. But limited vision definitely limits our freedom to build totally new concepts.

Can't get there from where? … from where we are.

In evolutionary terms, our currently popular species of town may functionally and economically be at a dead end.

It is unlikely that we will ever find a significantly improved vision by continuing the same processes and starting from where we are. We need to step way back in our thinking to a time before horses. We need towns designed simply for the needs of people.

We need to totally re-think what we are trying to do; we need a totally new set of parameters and ways to achieve our objectives.

All the interrelated systems in this huge complex industry do a magnificent job in what they currently coordinate and get done. They provide the possibility for more home ownership than ever before possible.

With a new vision, almost everyone would still fulfill similar functions. With a comprehensive new vision, however, the results could provide more satisfaction for everyone.

# Let's Just Walk?

*Americans like to walk, in the right circumstances*

Americans can walk—at least there's some encouraging evidence that they can. Americans are not anatomically connected to the automobile. Contrary to popular opinion, Americans of all types can walk, and they even seem to enjoy it, given the right circumstances.

Most will tell you, one of the pleasures of going to Europe is walking. They talk at length about the pleasures of strolling around cities and villages where cars have been eliminated from the streets. Especially interesting are narrow passageways winding up and around hillsides between beautiful old structures.

In European downtown shopping areas, they describe moving in and out of the continuous flow of people, any time of the day or night. They especially enjoy watching other people walk while they have a drink at an open-air café in some beautiful little pocket plaza. The atmosphere is often enhanced by intense beams of sunlight between buildings and penetrating into recesses of cool deep shadows, a refreshing retreat on a hot day. People are willing to pay a lot of money to go to such places and experience that. Some even stay.

Disney knew we were desperate for something similar to that kind of experience and the opportunity to share it with our children. Millions of us pay just for the privilege of walking around in such a fantasy. Is it fantasy or just a sampler of what our cities have lost? Is it simply the idea that a town environment can be fun? If people had to compete with cars on those Disneyland streets it would obviously be a distraction.

If we built a new concept town with real live pedestrian streets that were fun, how much would visitors pay for admission?

People flock to shopping malls and big-box store centers. They will put up with walking nearly a quarter of a mile in a hot or rainy parking lot. This gives them the chance to walk some more, looking at everything from one end to the other, once they're inside. If a mall is interesting enough, shoppers don't seem to notice how far it is to the next anchor store, especially if they can't see it.

People also flock to Las Vegas, by the thousands. Once there, cars are parked in parking garages bigger than some small towns and the visitors walk long distances to their hotels. In most places, it is necessary to walk a quarter mile for basics (e.g., going to the restroom). There are indoor imitation streets, with fake sky and clouds, just for people, … it's part of the entertainment. People we might least expect to enjoy the pedestrian life stay a week and never go near their cars. They go from one end of the strip to the other, walking and riding various forms of transportation.

It's impressive how many people use the monorails and trams. Perhaps Las Vegas is training increasing numbers of people to abandon their cars to enjoy the pleasures of walking and riding transport. This may be the first step in developing a market of home buyers interested in a town designed without cars.

Americans can walk. So why are developers and planners still so committed to providing for cars in currently popular housing and town solutions? Do they have any choice? Maybe not.

That's because the wrong challenges are given to them. The problem or design program given to them to solve is often limited by the past. Their imaginations aren't given a chance.

Wrong questions get wrong answers.

There seems to be no client with enough imagination and foresight to ask for a truly long-term solution for a complete town comprehensive

enough to address every human habitat requirement, be an integral part of nature, and, above all else, be designed primarily for people.

Once such a new concept town is understood and built, it will demonstrate how affordable, practical, and rewarding life in it can be. There will be no problem with designers, planners, environmentalist and developers working together, ... offering their hand at refining it or building something similar. That's because all the objectives of their accumulated principles of design and development will have an opportunity to be satisfied better than in any human habitat previously proposed.

*Part Three*

# People, Spaces, and Needs

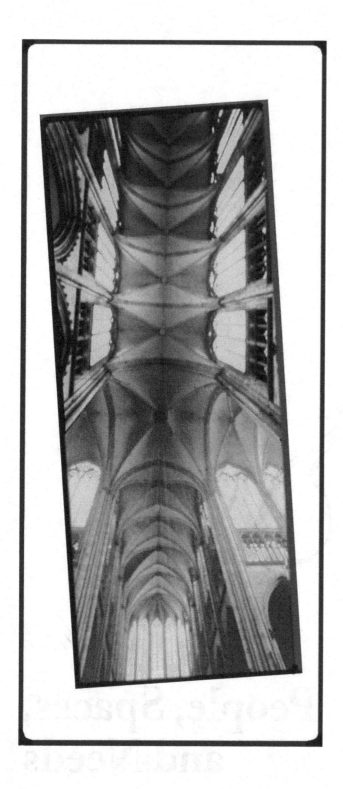

# Visualizing Something New

*Nothing new? New technology gets new results*

In architecture it's been said there is nothing new. Building design is often a matter of rearranging things from the past. But that's not the total story. Occasionally there is a new technology that generates an entirely new kind of structure. The flying buttress gave us the sun-lighted, stained-glass, 150-foot tall interiors and delicate stone work of Gothic cathedrals with 1000-foot spires as early the 12$^{th}$ century.

Imagine what it was like, visualizing and starting an entirely new building type that would take hundreds of years to build. Suddenly, proposing an entirely new concept for a town doesn't seem like such an impossible dream.

To introduce anything new to a society can be very difficult, especially something that's never existed before. Only those with exceptional visualization skills can fully imagine what it would be like to be in a new architectural space.

For anything new ever to be built requires a strong team of individuals who can each see the value of what is proposed. For those who fully appreciate the challenges and how things are built, a concept should sell itself, provided it's not so big or so complex that it's impossible to imagine. Before it's built, models and video presentations can help.

For the marketplace, sometimes a change in taste is required. We have seen significant changes in people's attitude toward music. This subject is not raised to discuss the matter of quality but rather simply to point out how the acceptance of something different, like sound, has been a gradual process. Video combined with music is a very pow-erful tool for change. It has been an effective medium for introducing many new things, even behavior or styles that previously might have been considered unacceptable. Of course, not all change is good.

Theoretically there are those who are trained to design new things. But even a talented designer's creativity can get lazy, can be trapped or led around by the cycles of style. That's often reinforced by trade publications.

It is comforting to think we have solved new problems by retreating to solutions from the past; people are even willing to overlook inherent shortcomings. But the current problems of our cities and advanced

technologies beg that we at least examine all the new possibilities available to us.

It's not commonly recognized that, for permanent buildings, an entirely new structural technology became possible in the 20th century.

Up until then, for thousands of years all durable buildings we humans built used only four basic systems of construction. Now, for the first time in all of history, structural fabrics give us a fifth structural system for permanent buildings.

In 1964 it took me several months to convince MIT we would some-day use fabric for buildings and that my research was worth while. After that I spent several years doing some of the original research in fabric structural systems. To the best of my knowledge, I designed and had built the first applications in the world on a department store, a church, an office building, and a residence. Notice the organic-looking shapes in the photos of those projects shown here.

The system uses a permanent structural fabric. The first structural fabric developed was made of woven fiberglass impregnated with Tef-lon. It's less than 1/16″ thick but it can span 150 feet without any inter-mediate support. As a building envelope, it is the structure as well as the interior and exterior finish; it can support the same roof load requirements as traditional rigid systems like steel; it's fireproof; it's

translucent, flexible, and does not appear to deteriorate. The weight required to enclose a space is less than a tenth of normal construction. The 6000-square-foot fabric roof for the church came in a 4x4x12-foot box and was lifted into place in one day. Computers allow us to design and build structural tension shapes that resist winds and weather as well as metal or concrete.

Almost all life forms on earth use membranes as their basic structural system. Every cell in every life form is enclosed in a membrane. Think of the infinitely beautiful shapes of membranes possible in a flower, jellyfish, or spider's web. The shape of a membrane is the pure expression of all the balanced forces within the structure. It is the minimal surface between all edges and supports. Human membrane structures are the closest we have come to a building type that clearly reflects the systems found in nature.

It took more than 10 years from my original research until the appropriate material was manufactured in large enough sizes and the public was ready to accept something so totally new.

Remembering those years gives me encouragement for this new town idea. It rekindles my patience and persistence and helps maintain my confidence in having chosen to accept the compelling urge to explore new dreams, to suggest to our culture that this process of dream-chasing itself is an interesting phenomenon worth exploring.

Fortunately, those many years in pursuit of membrane structures had a happy culmination.

By then, however, my interest had been captured by the very serious challenges posed by our inner cities and suburban sprawl. A new dream had been kindled. After having my work published internationally and recognized by *Time* for one the five best designs in the United States in 1981, it was a difficult decision to change direction. Nevertheless, I abandoned that newly blossoming structural fabric portion of my career, as well as a successful conventional architectural practice and several partnerships. Now an additional 20 years have passed, much longer than I ever expected. Researching the questions of what makes great cities or towns has been a wonderfully stimulating and challenging quest.

The greatest difficulty was avoiding the temptation to come to conclusions prematurely. It's good that I did not have a fixed date to conclude my research because each time I seemed to reach the solution, a new breakthrough in my thinking occurred that required the concept to evolve to a new level of refinement.

Only now am I finally ready to attempt to present this concept to the general public. I'm still open to and searching for new ideas. In my architectural practice, my staff often criticized me for making improvements and revising contract documents to a design right up until it went out for bid.

But I have found a solution to discovering new design ideas at the last minute. It's simply accepting change as one of the fundamental requirements for this town concept. Even after being built, it's designed to be flexible enough to accept and evolve with unexpected new needs and technology.

# Neighborhoods

*A pedestrian street, heart of neighborhood*

A street, especially a pedestrian street, can be the heart of an extended neighborhood.

There was a time and place when people sat on front porches in small towns. Everyone knew each other. People walked by and nodded greetings, and children played along the sidewalks. There was always enough going on to maintain interest and someone always enjoying the scene.

Something very similar occurs on streets in cities in many parts of the world. It could have been in a special part of Boston. Anytime, day or night, seniors sit in windows of four-story housing watching the street activities. Small groups, on benches or standing, visit in front of shops. Kids play, shoppers pass by, and occasionally someone calls out or waves a greeting. It's not unusual, even late at night, to hear a loud directive in a special tone of voice from a grandma sitting in an upper-level window, "Hey, Louie, stop that, or I'll tell your Mama!"

Imagine walking on a narrow cobblestone street in Denmark or winding up a hillside in Spain, with white plaster-walled houses on each side. It widens into a little square with kids playing, laughing, and dashing around. Two ladies standing off to the side in open doorways watch and nod as we pass by.

Late in the day, friends in a hill town in Italy stop to visit at a terrace on the way home. Some hold small packages. They pause to admire the sunset and reflections off a small pond at a distance in the orchard below. Others step out of doorways to join the group.

Through a Scottish close, past shops facing the street, you enter a courtyard. It is surrounded on all sides by four or more stories of housing. This is a multi-use space, hundreds of years old, for unloading horse carriages in the past and cars today, then both continue on to garages beyond. At the moment it is for neighbors visiting and kids playing.

You can drive or you can walk under beautiful trees hanging over the street, past manicured lawns and stately homes for blocks and blocks. Occasionally cars come and go from behind automatic garage doors. There's not a person in sight.

Other neighborhoods have no trees, long streets, bumper-to-bumper cars along the curbs, a few open hoods, deteriorated homes with broken windows, and a few people who stop to stare at anyone driving by.

The word "neighborhood" has to stretch to cover many very different kinds of places. The first groups describe characteristics we can achieve; the last two paragraphs mention problems a new town concept can solve.

Great neighborhoods and their activities, … what are the physical characteristics that will allow a neighborhood to simply blossom as part of daily life? We can recognize the important features. It's easier for people to get to know each other if they encounter each other on a regular basis.

Anyone who has lived in a high-rise apartment building with several apartments per floor knows that rarely do people see each other in an interior hallway walking to the elevator. The walk is a matter of seconds and there are many seconds in the day. A person walking from a house on a street with several houses can be seen by any one looking out a window. Open balconies of apartments around a courtyard are similar. Simply being seen within a distance to be easily recognized or, better, close enough for waving—both encourage greetings and getting to know people.

There can be nodes of activity along the way: kids playing, a beautiful view, artwork, a magazine rack, a shelf for things people want to give away, a basket of fresh fruit, a kid's lemonade stand, maybe grandma's cookies, coffee, or anything useful that gives people a reason to pause for a few moments. Add to any of these a few chairs and a table, then encounters can be multiplied many times. Things like this can be pulled together and refined from our past and from other cultures. Freedom can be provided for designs to vary depending on local desires.

The arrangement of spaces is important. The approximately 100-year-old department store in downtown Philadelphia has a grand interior space and several floors, almost like front porches, overlooking its street level floor. This and John Portman's design for Hyatt Hotel's original atrium in Atlanta, or the interiors of Embassy Suites hotels, begin to hint at possibilities. Naturally the design character and scale could be considerably different. Something on the scale of

the insides of the larger European cathedrals might be the size for an extended neighborhood.

There seems to be no existing space or neighborhood that does it all, but here are some basic concepts for our housing market: grouping houses so that a relationship between them is clearly implied; no more than ten or twenty homes; front porches on all homes, close enough to wave greetings. As an extension of the home, the porch is the living space used to let visitors know they're welcome. A play area where parents or seniors can share watching over young children playing should be visible from each home. It will take people living there a while to add special features. The walkway from each home to everything else should pass the play area, which is also a good place for visiting and forming friendships.

This cluster could be arranged so every porch and the play area overlook the plaza and sidewalk cafés on the shopping street down and off to the side. With other clusters overlooking in the same manner, an extended neighborhood of clusters could be created, with their local section of the main shopping street as the center.

The concentration of housing related to each local section of the shopping street is necessary to guarantee its success. This is important. Several residential streets are needed to support one commercial street. If the homes on all of those streets can have a direct positive connection to that same local section of main street, both are enhanced. This could be simply visual, watching friends on their way to other activities. When in separate locations, each of those several additional streets lack enough traffic for vitality, are a maintenance burden, and can dilute pedestrian traffic on the shopping street.

This essay outlines a few factors affecting an extended neighborhood. The challenge is to combine these positive features with all the other comprehensive objectives of an environmentally integrated town.

# Types of Housing

*Connections add importance to the "people community"*

Through many stages of life, our home is our most important place. Housing is the basic building block of a town. Currently, streets rather than homes still seem to dominate as the basic form-giver.

The way the physical home is connected to the immediate neighborhood and the rest of the town can significantly influence how important the "people community" becomes to each of us. A lack of connection leads to anonymity, which contributes to increasingly antisocial behavior. On a human level, our current towns lack almost any connection or convenient transition from each home to the various levels of public space. That creates isolation, especially for the lives of children and seniors. They're trapped because another person's time is required to drive when they can't drive themselves.

Often, another lack in suburbia and cities is any significant connection to natural open spaces. People are engulfed in totally man-made surroundings. The bigger the city, the more unnaturally alien can be the spaces and the lives possible in them.

Of course, we must remember that not all cities or suburbia are totally bad, at least in the beginning. There are things we like about them that will be hard to fully duplicate. What we need to discover is a balance of the best features so that the total result is better.

Our current approach to housing treats us a little like pigeons; if you want to live in a certain housing type, you must build in a certain zone. Zoning is designed to protect adjacent properties, but it also causes many of our problems.

Better ways are needed to protect property values while also allowing greater freedom within each property. Home sites should be able to accommodate a single-family residence, duplex, fourplex, or even a small family restaurant or shop without detracting from the neighborhood. That is much easier if cars are not a factor.

Affordable housing is usually built as a special segregated neighborhood. Often a stigma becomes identified with that area and its occupants. A need for affordable housing can come at different stages of life. Being young, old, or having difficult times means having and needing less. Almost anyone can have periods of time when they need less space, and it would be nice to have the freedom to retreat into fewer rooms.

A young couple or a senior could supplement income by renting out unneeded space. Greater flexibility in housing can help maintain neighborhoods and life-long friendships, … for many generations.

Any new town concept will need the full variety of all housing types. Reviewing some factors we know about existing housing may be helpful as we seek new ideas. The examples below touch on a few basic features of our most common housing types.

## Tract House Developments

This is the dream of most people, at least in Western states. When lending rates are low, builders cannot build them fast enough to meet the market demand. Most critics see this as the most significant contributor to traffic and farmland consumption.

During the 1950s in the Eastern states, homes tended to be two stories. The one-story Western ranch-style homes had wider lots and fences. Today in the West, an average tract home site has a buildable area about 35′ x 55′, with a height limit of two stories and roof to 35′. Houses are built within the limits of that imaginary three-dimensional box. House sizes range from 1250 to 2500 square feet or more. Most have a connected garage. Side yards provide from 6 to 10 feet of separation from neighbors' windows. Most back yards have no views and neighbor houses overlook the back fences. Front windows view

required front yards and parked cars along a street with sidewalks. Initial construction and land costs are as low as possible for a particular market. Long-term maintenance costs can be expensive.

Trees eventually add character and are pleasant to drive past. Occasionally children play in front yards or pass by on their bikes. Often on weekends people work in front yards or wash cars. But the separation between houses is usually just beyond easy conversation distance and there is seldom an excuse to get close enough to visit. Of course, all this varies depending on the neighborhood. Getting together is usually by invitation, if neighbors make that special effort.

Here's a bit of cheer for the prolific critics who truly dislike suburban tracts. Suburbia may have one great advantage over most other housing types. In the future they may provide land for a better idea. After a hundred years or so, as the infrastructure and maintenance cost become unreasonably high, they will be more easily bulldozed.

## Condos and Planned Unit Developments

These types are often closer to work, shopping, or transportation and this housing type is very common in New (Traditional) Urbanism, where housing is often over stores. The investment is usually smaller than that for a tract house, but the maintenance costs are not in direct control of the individual owners.

Many have a patio or balcony that the individual maintains. Cars can be parked in attached garages, shared basement garages, or parking lots or on the street. One end of the unit usually views a street, semi-public access courtyard or walkway. Often the other end, typically bedrooms, overlooks driveways or parking. Some have back yards.

Condo-urbia is usually between suburban tracts and downtown. They usually have standard floor plans and recreational amenities. Projects range from one story to multistory, some with shops on the first floor. In the long term, if not done to the highest standards, large areas of these developments could be just another form of suburbia, just taller and a little harder to remove.

Socially there seems to be slightly more neighborhood life than in tracts. That may be more a result of a younger generation than the design. The closeness of units to each other and people passing in the walkways or courtyards may cause more interaction. More frequent turn over in ownership, however, may offset this.

### Loft Condos

This type has become very popular among younger buyers. Often they have the freedom to do the interiors the way they want and save cost. Old industrial buildings are often converted into lofts. Currently, new 10- to 15-story single-building loft developments in downtown areas are very successful. Most have two-story loft spaces with full-height glass walls and patios overlooking the city.

### Rental Developments

Apartments are a big segment of the housing market. They range from an owner occupied duplex or fourplex to 500 or 1000 unit apartment complexes. Floor plans include the entire range of housing types, but the largest number are probably two bedrooms with bath, living/dining, and kitchen. Entrances are usually off a double-loaded common hall or a single-loaded exterior walk.

Two to four stories are typical. Depending on local requirements, the height and type of construction are often limited by fire-fighting equipment. Many have extensive amenities. Of course, every other form of housing can also be rented.

## Other Factors

Around the world, buildings come in many forms and variations based on many factors. It doesn't seem to matter what the design concept may be; once building design patterns are established, they are often built on available land as far as the eye can see — not because it was planned, it just happens.

Natural surroundings usually improve livability and value. Various cultures develop their own unique design techniques for reflecting and relating to their natural setting. Often they try to become part of their surroundings, but some cultures ignore their surroundings.

## Other Uses

In many cities, especially in older sections, other uses found to be compatible are permitted in residential districts. The freedom to have a variety of uses should exist in any new concept expected to last hundreds of years. There will probably be new types of uses that we are not even aware of yet.

A family might have a four-table restaurant, a beauty shop, an accounting service, etc. This sort of activity will be practical only if cars are not allowed. Pedestrians coming and going will have little effect on neighbors. Naturally it would be good to have some sort of neighbor approval process. It will also be influenced by the location of the immediate neighborhood and its relation to Main Street.

## A New Kind of Home Site

Objectives for a totally new kind of house ideally should allow flexibility for all of the types of housing and features mentioned above. Naturally all may not be possible, so the attempt should be to discover the best set of combinations.

The approach to achieve these objectives is to invent a new kind of *home site*. The essence of many of the current problems comes from the restrictions on the site. Those restrictions are necessary when all sides of what's built are visible to neighbors in order to protect neighboring property values.

If a new kind of site can be developed that gives maximum protection to or from the neighbors and which also gives greater freedom within the site, … then most of the needs and flexibility objectives of each home will be solved.

# Other Countries
*Centuries ahead in population-to-land ratio*

There are countries in the world that are centuries ahead of us in terms of *population relative to land area* available for housing. Many in the Far East have thousands of years of experience. If we don't plan for such a long future, we may not have one. While many Americans currently feel such conditions in other countries have no relation to us, each offers perspectives on future possibilities we cannot afford to ignore.

One fascinating example is South Korea. This is an approximate description of several housing-related matters as I recall them. They may not be 100% accurate but will give the general idea.

Much of the South's housing was destroyed during the struggle against Communist efforts to dominate it. South Korea quickly rebuilt five-story housing on most of the buildable land.

Because of ancient beliefs there was a limited portion of land available for buildings; the rest was for farming or open space. With their newly found political and economic freedoms the lives of individuals quickly improved, and soon there was a demand for more and better housing. Within 25 years, almost all of the five-story housing had been removed and replaced with 12- to 20-story housing.

Now those are beginning to be replaced with the first 60-story towers. That may be the future, as long as limits on land for building

remain. Those first towers, for economic viability, were combinations of hotels, offices, or housing.

Individual home layouts have a shared entry with doors into two or more living units for the related families or sub-tenants. Some share living areas. The concept of government-subsidized housing does not exist. Families work to help each other. The plans and exterior walls of the house can be changed to suit the family's needs. This does not affect the exterior appearance of the building because there is a four-foot wide continuous solarium is on all sides. The exterior is all glass and that is what's seen from the outside. The logic is, (this is my interpretation because I have also done it for energy-efficiency) if you're going to have dual glass, you may as well use the space between the two pieces of glass and have an energy-saving solarium as well a considerable amount of specialized usable space.

As a standard practice, families interested in being tenants must deposit money in the landlord's account, enough so the interest pays the rent. It is refundable if you move. Housing towers are largely tenant-financed before construction starts. An initial deposit is made, and as the project progresses additional deposits are made until completion and move-in.

New towns are a series of 20-story (or higher) towers, including basic commercial, service, transit, schools, and recreational facilities. Unless they are built in an existing city, they overlook flat farmland valleys and tree-covered hillsides. The new town populations are in the tens of thousands.

If a society is serious about limits on land for building, at some point high-rise housing may be necessary to meet future housing demand. Sound long-term planning could avoid the extra expense of removal and replacement cost.

# The Ideal Home
*People like country and city life, ... choose both*

A good location can add value to any house. Many people will pay extra for a home in the country on a hillside with a good view. Others find more enjoyment in life downtown with convenience and vitality. Most people would like both, but are forced to choose one over the other.

Each home needs areas that are totally private. A private backyard would be an improvement over what is usually available in tracts, with neighbors overlooking fences. Each home needs visual connection to meaningful spaces outside the home; a dream would be views of the natural countryside.

Front rooms and porches need to be arranged to encourage interaction with the neighborhood, but interior rooms should be totally sound-proofed from neighbors on each side. Acoustical science shows the only way to achieve that is with a solid wall.

People are forced to move from their favorite neighborhoods because their house no longer meets their family needs. Relationships and neighborhoods are weakened. Each move has a loss in escrow, finance, and real estate fees, and property taxes go up with any sale. None of this adds to the physical value of the ownership investment. Many reasons for moving can be eliminated by making the home flexible enough to adjust as the owner's needs change.

For many people, their home is often their largest and best investment. Homes should also provide more opportunity for sweat equity as a method to increase a family's financial security. If the owners could easily add a room themselves, that saves paying someone else with money that's been reduced by various taxes. Direct purchase of materials could save fees. If done gradually as savings are accumulated for materials, there's no need for a loan, so savings on interest can result in 30% to 50% more construction for the same dollars. Sweat equity can provide the most value for time invested, especially if done between jobs (i.e., when unemployed).

With the freedom of easy redesign, the potential for the home as an investment should be significantly improved. A home can become the stepping-stone to a financially improved future; it could provide a sound investment and rental income for old age.

The housing and town equivalent of a quantum leap is necessary to comprehensively address all the long-term sustainability and human issues. Nothing currently being built does that. Developing the techniques to meet these challenges is the objective of the concept presented in this book.

Something significantly different is hard to imagine living in. The first prototype will have to be attractive enough to pull people away from their tract home dream.

## Tract Home

The typical tract home would be the primary marketplace competition for homes in a new town concept.

Briefly, for comparison, driving down the street in a typical tract development, the homes vary a little but mostly look the same. Cars and garages are a major element in the street scene. There may be an occasional person in a front yard next door or across the street, but usually not. On each side of the house is a long narrow slot with neighbor's windows a few feet away. The typical front door often has an entry stoop; some may have large columns supporting a higher portion of the roof for appearance rather than any practical use. The interior is for current needs but it often has spaces that are seldom used, perhaps only during holidays. One of the gathering-type rooms next to the kitchen opens onto a back-yard patio. There is no actual view. The backyard is fenced, with the neighbor's second floor windows overlooking on three sides. In parts of the East the back yard may be open, facing the backs of similar houses.

## Dream Home

By contrast, imagine we are in the future and have discovered this fabulous community that's only three hundred years old and we are just entering the immediate neighborhood of our imaginary ideal home.

It's in a small cluster of homes with each front porch overlooking a play area. We are in the play area. Children are motioning to a friend on an upper porch to come down to play. Three grandmothers are looking after the younger kids. They've known each other since they were kids in this same neighborhood.

As we pass under a handsome structural arch, our agent waves to a couple of his friends on their porch. We are walking along at the level of treetops (growing in the plaza below) bathed in bright sunshine. The tile walkway is like a terrace with a combination of massive limestone guardrails and sections of clear glass with views to activities on the street below. Two couples are playing an intense game of 3-D holographic dominos on a porch. They nod as we pass by.

Another couple is standing at the guardrail signaling to friends seated at a café a short distance off on the lower street level terrace. They turn to greet us, saying how much they have enjoyed living in this cluster of homes and the rest of the community.

The air is pleasant, with a hint of orange blossoms, said to bloom here at many times throughout the year. Flowering vines growing 20 feet high spread across the 10-foot-wide overhead surface that gives the effect of sheltering the walkway and porch. We pass under several delicate trees with wispy branches of bright green lacy leaves.

The front of the house is simple, with large arched windows and a 15-foot-high wood timber and tile roof over the porch. The porch is 10′ x 15′. The front design of the home is easily changed to our taste, but it may need the neighbors' review.

We can see the neighborhood restaurant, almost next door on the terrace below. That's the place where most people hold big gatherings for family and friends. We are told the owners are like part of the family anyway. Open-air neighborhood picnics are held in the play area.

The room interiors can be changed almost as easily as moving furniture; floor and wall panels simply snap together. Additional panels are available by delivery from Home Depot.

There is a utility room with an elevator directly to the home's private two-car garage on a lower level. Most of it is used for storage because only one car is necessary and it's much smaller than those when the garage was originally built.

The back of the home has the famous views, a trademark of this town. The backyard also has a large patio, hot tub, garden, and lawn area with a swing.

The town's permanent ageless concept and mature landscaping provide the opportunity for personal touches to enrich its character, *over hundreds of years.*

Retuning our thoughts to today, what we begin to realize is that such a home site would have an exceptional location and all the flexibility to easily change the home any way desired. But how can so many exceptional features be possible?

# Diagrams of Basic Needs
*Help in visualizing desired objectives*

Basic diagrams may help to visualize arrangements that can make the most important objectives possible.

An individual home site needs to be a three-dimensional volume, connected to the rest of the community by a common area for circulation. The pedestrian and visual link is by way of the front rooms and the porch. These view the other homes in the immediate neighborhood, a play area, the extended neighborhood, and the shopping street beyond.

The opposite end of the home site has the private backyard. The back rooms and the yard need views of the open countryside, with no other homes in sight.

The heart of the home is the totally flexible and private living space in the center of the site. The interior spaces can be arranged to meet almost every imaginable need.

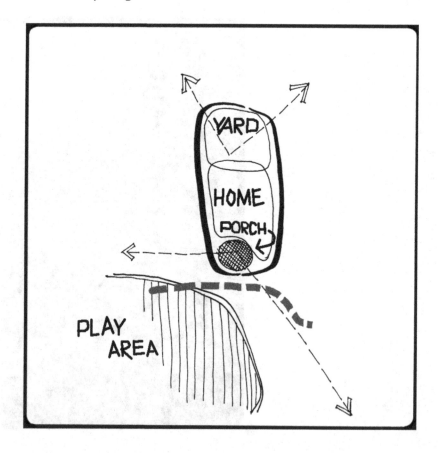

Home sites need to be clustered around one side of a play area. This will form the immediate neighborhood; this is the basic Cluster. All of these overlook Main Street. The play area may also extend to the countryside edge of the Cluster, where it could have a swimming pool.

Homes relate to their play area as if it is their ground level, allowing the Cluster to be elevated, as on a terrace.

Clusters may be stacked. To create an extended neighborhood, another stacked group of Clusters are placed on the other side of the street with their play areas also overlooking the same section of the street.

This section of Main Street becomes an integral part of this extended neighborhood.

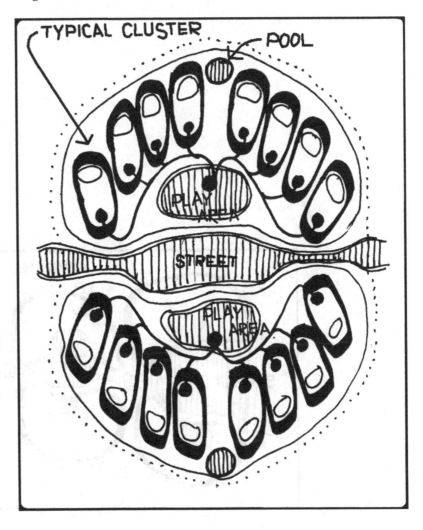

Main Street connects all the Extended Neighborhoods to each other; it becomes the heart of the town. It's the circulation path to everything.

The density of housing gives livability and vitality to the street and to the shops so they can afford to provide for the needs of the people. The continuation of Main Street, partially enclosed on both sides by the stacked Clusters, forms a town that can adjust in shape to fit the contours of the land. Openings between Clusters give the street vistas of the surrounding country side.

Main Street can be open to the sky or closed as necessary for the climate. It can vary in width to meet community needs for larger public spaces.

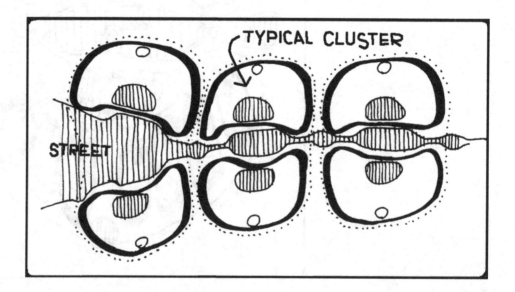

# Would There Ever be a Client?

*To think so comprehensively or justify the research?*

The necessity for doing this research became clear many years ago, when I finally realized *there would probably never be a client who would recognize the total challenge—the necessity or opportunity to comprehensively change the entire way we do cities.* No one is trained or likely to imagine being a client who would seek totally new sets of arrangements for a new kind of town that would provide optimum opportunity and flexibility for every individual, home, and neighborhood—a town designed to look and function as an integral part of nature, optimum for every individual inhabitant in our time and durable enough for hundreds of years in the future, with individual home prices about the same as for a quality suburban home with comparable amenities—and then to combine all this with all the other complex interrelated challenges facing a town.

There is probably no developer or government with the imagination, foresight, budget, or desire to even think in such comprehensive terms, not to mention justify the years of research. And even if there was one, it would only be a facilitator. Ultimately such a client would not be the end user, … not the real client. Personal objectives and preconceived thinking could become a greater influence on the concept than any other single element. It's probably fortunate a typical client didn't exist.

It was necessary to invent an ideal client, that entity ultimately responsible for using, measuring, testing, and evaluating every design decision.

Over 25 years in my conventional practice I had the opportunity to design for many great clients. Some examples are shown here. My most original and cost-effective work happened when budget or other constraints forced me to seek non-conventional alternatives, … even to imagine the client. These resulted in the first applications

of silicone rubber to attach and hinge glass, below-grade low energy housing and offices, low-cost housing at one-quarter of tract house cost, structural fabric systems, cooling effects of double metal roof systems, berm-wall construction, 12-story concrete block housing, solarium-heated apartments, roof-truss attic room additions, sweat equity housing systems, and developing a way to continue these years of research, … free from any limiting outside influence. Each of these is a story of successful but little known, marketplace solutions that built confidence for tackling bigger challenges.

111

That was the starting point, so for this town's research it was necessary to imagine the most influential client, one who will care most, … and will be most influenced by home and town; … it has to be none other than each and every individual, within their personal community and culture, at every stage of life, doing every imaginable activity, and activities not yet imagined, … plus the planet we currently live on.

*Part Four*

# Habitat Solutions

# The Design and Planning Process
*Home, only knowable part of a never-before town concept*

To design a town, a city planner would never start by first designing one of its many individual homes. But that's the only starting place for a never-before town concept.

Normally, considering plans for a complete town down to an individual house involves at least four different specialties. Planning focuses on overall circulation, use patterns, laws, and ordinances. Urban Design is about the arrangement of large buildings and spaces often visualized as giant landscape elements. Architects do buildings; trophy buildings that contrast with the surrounding context often get the most attention. The Builder's team usually designs houses.

However, if starting from scratch, the individual home is the only knowable element, the basic building block, and even it needs to be reinvented. Then patterns of homes make neighborhoods and neighborhoods make the town. This approach blurs the areas of different specialties. Trophy buildings, arrangements of buildings, and circulation patterns inherently represent preconceived approaches. The same types of spaces are needed, but with a new format of integration.

The next three paragraphs briefly review town issues already discussed. Typical current towns can be thought of as large structures spread out on two-dimensional street patterns. It seems obvious spreading all the pieces spaced apart on the ground is not the most efficient system for reducing energy use, utility runs, construction costs, or land used. Compared with the world's favorite historic villages, our towns seem like wasteful primitive diagrams. New knowledge allows new village designs.

Compact 3-D arrangement of spaces allows minimum land use, construction costs, and utility runs but maximum privacy and open space. We see this principle in the cellular structure of a flower and from coral to honeycombs. Each cell or occupant has private space.

The individual homeowners should be able to build whatever they want, with more freedom than in typical 2-D street diagrams. It would be wrong to have any preconceived notion of what a town should look like. The entire concept must grow within a basic infrastructure of services and the functional design needs of the individual home. The process may be related to fractals, but not as a decision-

114

making guide, because towns have too many variables of continually changing needs at every level of scale.

Anasazi cliff dwellings offer clues for settlements built primarily for people. They were very compact multistory structures and had rooftop terraces with views focused on common areas. Mesa Verde is built way back under a long flat rock shelf. At Betatakin (next page) the homes are in a magnificent rock pocket over one hundred feet high sloping up and over the front edge of the settlement cluster. Looking outward from way back in the cave you don't even notice the upper edge of the opening or realize a huge rock formation is hanging over you. The forward locations of the lower side edges of the arched cave provide a dramatic sense of shelter. But these only give us a few more clues and a little encouragement.

Our community design and planning approach needs complete freedom in order to invent how everything we discover can be combined to meet all our future objectives. This section of essays explores how to design the basic people spaces and the concepts for meeting people's needs, ... allowing them to live to their maximum potential.

# Home Site

*Affordability, sweat equity, improves family financial futures*

The challenge is to invent a residential lot, a subdivided space, that can be purchased. Provide enough other outstanding features to attract buyers from all other home site choices.

## These Are the Basic Design Requirements:

Location, Location, Location:
- it has to be in an ideal location, offering backyards with near total privacy,

117

- like on a hillside in the country with no apparent neighbors and a beautiful view,
- back yards may have personal items such as bird houses, waterfalls, and connecting gates,

- each back yard must allow complete freedom of design to suit the homeowner's desires — any style, including very traditional, modern, or very natural,

- two minutes' walk to the local hang out, such as a sidewalk coffee shop with a newsstand, also visible from porches,
- each home an integral part of a town, an easy five-minute walk to everything, e.g., shops, recreation, services, transit, schools, etc.,

- with front rooms and porch overlooking an exciting shopping street and the fascinating activities along the fronts of the many other homes in the extended neighborhood, all within waving distance,
- each needs close neighbors clustered around an area visible from each home site for greeting, gathering, and children playing,
- and no cars in the people spaces.

121

But there shall be complete design freedom to suit individual taste. Different back-yard and front-porch designs are shown as possible examples.

Each lot must allow flexibility for any floor plan, for every stage of life, adjusted to meet any budget or needs, and easy for owners to modify themselves. It should allow for two-story construction, the highest energy efficiency, and easy access to a private garage.

It should allow for a home size from 400 square feet to 2,700 square feet that can be added to or changed at any time without disturbing the neighbors. Highest-quality soundproof separations between neighbors will be achieved by solid construction and by the arrangement of spaces and service access.

Each site shall permit a single family residence or duplex, tri-plex, or fourplex units of various sizes as desired. Each lot shall permit a home business, artist studio, small family restaurant, or shop. Since long-term future use is difficult to predict, with community approval home sites should also accommodate any use from office to light industry.

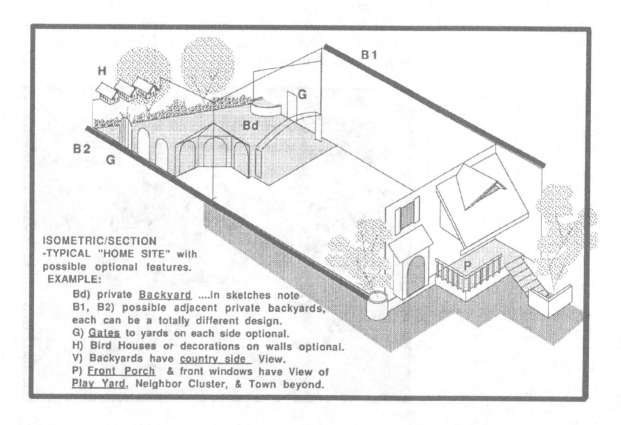

ISOMETRIC/SECTION
-TYPICAL "HOME SITE" with possible optional features.
EXAMPLE:
    Bd) private Backyard ....in sketches note
    B1, B2) possible adjacent private backyards, each can be a totally different design.
    G) Gates to yards on each side optional.
    H) Bird Houses or decorations on walls optional.
    V) Backyards have country side View.
    P) Front Porch & front windows have View of Play Yard, Neighbor Cluster, & Town beyond.

### How It Can Be Done:

The home site diagrams show the basic arrangement, with front porch, soundproof side walls, the outer edge of the back yard open to the view, and side fence walls sloping up to full height to give privacy to the entire back of the house. The site has a buildable height of 20 feet. Only the front wall can be seen by neighbors. Therefore, the inside of the site and the back wall can be whatever the owner wants. Of course interiors will need to meet basic health, safety, and welfare requirements. A private garage on a lower level is reached by an elevator, which is also used to bring supplies. A container box with major construction supplies could be delivered by crane to the backyard.

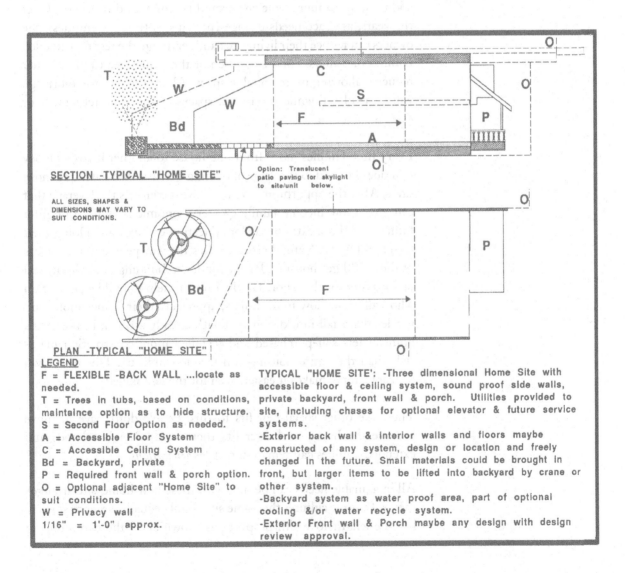

SECTION -TYPICAL "HOME SITE"

Option: Translucent patio paving for skylight to site/unit below.

ALL SIZES, SHAPES & DIMENSIONS MAY VARY TO SUIT CONDITIONS.

PLAN -TYPICAL "HOME SITE"

LEGEND

F = FLEXIBLE -BACK WALL ...locate as needed.
T = Trees in tubs, based on conditions, maintaince option as to hide structure.
S = Second Floor Option as needed.
A = Accessible Floor System
C = Accessible Ceiling System
Bd = Backyard, private
P = Required front wall & porch option.
O = Optional adjacent "Home Site" to suit conditions.
W = Privacy wall
1/16" = 1'-0" approx.

TYPICAL "HOME SITE": -Three dimensional Home Site with accessible floor & ceiling system, sound proof side walls, private backyard, front wall & porch. Utilities provided to site, including chases for optional elevator & future service systems.
-Exterior back wall & interior walls and floors maybe constructed of any system, design or location and freely changed in the future. Small materials could be brought in front, but larger items to be lifted into backyard by crane or other system.
-Backyard system as water proof area, part of optional cooling &/or water recycle system.
-Exterior Front wall & Porch maybe any design with design review approval.

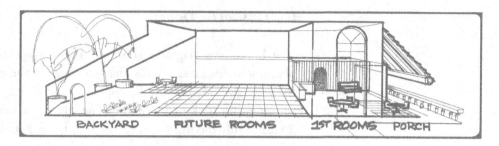

BACKYARD    FUTURE ROOMS    1ST ROOMS    PORCH

A young buyer can simply build a single low-cost living space with a bath. The remaining space can be yard or loft. This can make entry-level housing affordable, but here it has a tremendous up-side potential. As needs and budget grow, so can the home, ultimately to a 2700-square-foot home with several rooms. The subdivision of spaces for each home site can be any size best suited to the local market. In addition, up to four home sites could be connected if desired. Over the years, as space needs change (perhaps as the nest empties), the owners can reduce their living area and rearrange the rest for extended families or renters. For a single senior the living area can be further reduced, allowing more rentable space. Along the way the mortgage will be paid off, while the rented spaces can provide retirement or extra income.

Every time families move from one house to another it costs money that does not add to the value of owning a house. Not having to move saves. Also, the opportunity to invest sweat equity is the leverage that can give a low-income family greater economic stability, a brighter future, and those extra funds for education or investing. Doing their own building is a natural process — it's the way people in most parts of the world get housing. It's a confidence-building and educational opportunity for the whole family. It could help to build a generation who will know how to do things, appreciate their homes more, and not let them fall into disrepair. That's already a trend; in the 1990s homeowners self-purchased and installed more construction materials than the entire housing construction industry. This new town concept simply makes it easier, even for the less skilled.

There are other social benefits to this. People don't have to move because the house no longer fits their needs. Neighborhoods and extended families can remain intact *for generations*.

All imaginable needs can be met with these new home-site arrangements. In some cultures the home site could be made larger or smaller, but the length-to-width proportions shown allow the two rooms at front and back to have windows, with a small area in the center for storage, service, and equipment.

Technical details, energy efficiency and construction systems will be described in a following section.

All homes built today should allow for changes in size, arrangement, and affordability. Retirees, singles, couples with no children, singles with children, extended families, and various combinations may vary in percentages over time. This is not possible in other current housing choices.

Only homes built in this new concept town offer the comprehensive range of benefits that will make every condition and stage of life all it can be.

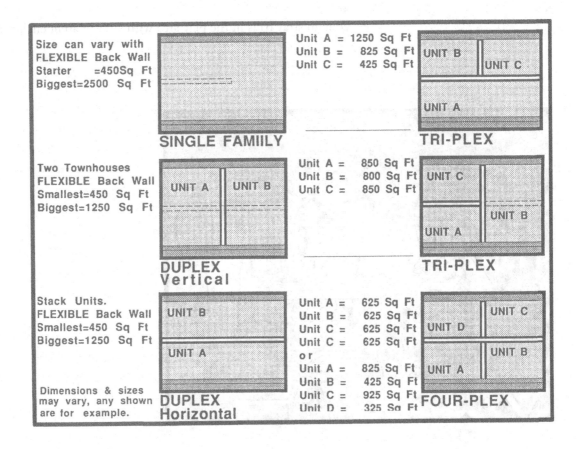

# Near Neighborhood
*Encourages livability, friendships, and community*

The goal is to develop a grouping of homes with a spatial arrangement that encourages maximum livability and formation of friendships. Provide enough other outstanding features to attract buyers from all other neighborhood choices.

## These Are the Basic Design Requirements:
- arrange homes to focus on a safe neighborhood area where children play and can be seen from each home.
- arrange all homes as close as possible to the play area.
- design each home so it can also view the street, areas of local interest, and the countryside.
- determine the best number of homes for this small neighborhood.
- provide for a possible local swimming pool in direct sun and where its noise would not be a problem.
- provide a walkway to connect each home to the common play area.
- contain stray balls in the play area with a transparent enclosure.

## How It Can Be Done:

The Near Neighbor Node has as its hub an open area of about 40'x80' (see diagram on next page). Aligned around one side, a row of home sites extends in two directions. Each home site offsets 10 to 20 feet (16 feet shown in the diagram) to give good views of the play area. This forms a Cluster. Note: 50% of the homes have three exterior walls for windows.

Each home site and local activities focus on this center area. It is on the most convenient pathway to everything. It is the node of activity, … greeting, gathering, and kids playing. There is space for tables, chairs, benches, or swings and a good play surface. It's great for games and street dances. It gets its abbreviated name, "Nana Node" from grandmothers baby-sitting younger generations.

The architectural character or designs shown are only to help in visualizing the spaces. The potential design variations are as unlimited as buildings on any site, as long as the arrangement is maintained within basic variations.

Four home sites extend from each side of the play area, establishing a comfortable viewing distance of about 100 feet. Homes are in pairs for sharing the elevator and utility chases. An upper level of home sites is placed above the homes on the play area level. This maintains

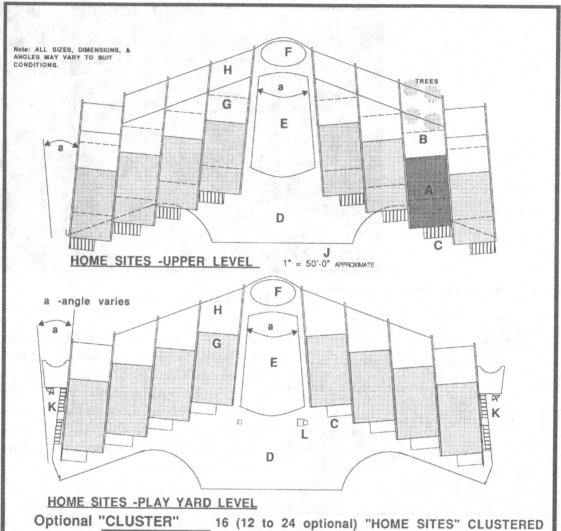

Note: ALL SIZES, DIMENSIONS, & ANGLES MAY VARY TO SUIT CONDITIONS.

**HOME SITES -UPPER LEVEL**    1" = 50'-0" APPROXIMATE

a -angle varies

**HOME SITES -PLAY YARD LEVEL**

## Optional "CLUSTER"    16 (12 to 24 optional) "HOME SITES" CLUSTERED around PLAY YARD & Town View beyond.

(A) HOME SITE. (for example, typical) -Interior arrangement flexible. Porch, Play yard, Town & Countryside VIEW.
(B) BACK YARD for Garden & Play. Private. Countryside View. Trees to hide structurres. Exterior wall any design.
(C) FRONT PORCH, -Play Yard & Mall or Main Street View.
(D) PLAY YARD, -Mall or Main Street, Porches & Countryside View. With glass guard rail at height to suit conditions and contain cooled air.
(E) OPEN To Play Yard Below.
(F) SWIMMING POOL or DECK AREA option (enclosure is optional) -Countryside View.
(G) Line of HOME Site on Play Yard level.
(H) YARD of Home Site on Play Yard level Site.
(J) Line of PLAY YARD & WALK WAY to Porches on Play Yard level.
(K) Stairway to LAKE, FOREST, ETC. Between CLUSTER options, Beautiful
     spaces for shops, etc & enclosed to suit conditions. Views to Interior & Exterior spaces possible.
(L) Guest ELEVATOR, From mall Level to each Play Yard level. With security access option.

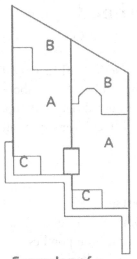

**Examples of Plan & Walk Variations**

comfortable viewing distances for a total of 16 homes and gives the upper level a better viewing angle. The overall long dimension can be thought of as a block, but it is shorter than a typical city block. Twelve to 20 homes seem to be about right.

The upper-level homes are accessed by a stair or elevator. A walkway passes in front of the play-level porches, continuing through play area, elevator, and then to Main Street beyond.

Everything behind the front wall of each home is totally private. The porch is an inviting extension of the living space. Some may have kitchen windows opening onto the porch. Here, as in many cultures, you can have backyard barbecue-type get-togethers on the front porch.

Where home sites have more than one unit, the additional units can also have porches or balconies. Even from the extended neighborhood across the street court, greetings can be exchanged.

All ages of people can live in this neighborhood, ... old or young, in small units or big units, and big families or small families. It is perfect for extended families, in the same home site or adjacent sites.

Not having cars in front of or near each home is the single most important reason neighborhood density can be increased. We often ignore the huge impact cars have on our neighborhoods. Pedestrians coming and going have very little negative impact on a neighborhood.

# Extended Neighborhood

*Convenient for greeting friends, safe for any age*

The goal is to develop a grouping of Nana Nodes with a spatial arrangement that encourages a sense of the extended neighborhood and facilitates formation of friendships from this larger group. Provide enough other outstanding features to attract buyers from all other extended neighborhood choices.

### These Are the Basic Design Requirements:

- arrange spaces to have enough things going on within view so people find it interesting and fun to sit on their front porches.
- increase security by friends looking out for each other.
- gather enough Nana Nodes together to financially support a block of shops, a local sidewalk café, or other places to visit at any time.
- make access to the street convenient to encourage participation.
- create an arrangement that brings together front porches of the extended neighborhood to within a distance convenient for watching or waving to friends.
- arrange so every home visually participates in its extended neighborhood.
- provide an exposure for young children so they can begin to realize they are part of their neighborhoods and the community.

### How It Can Be Done:

First it is necessary to reinforce our understanding or perception of "ground level." If you are in your backyard with a patio, lawn and

garden, you can't see what is under it, whether it is at sea level or on the side of a hill, even if the hill is the roof of a house below it. You will still come to think of it as your yard. For all practical purposes, it becomes your ground level.

THE SITE PLAN CONCEPT MAY INCLUDE BUT IS NOT LIMITED TO THE OPTIONS SHOWN HERE.

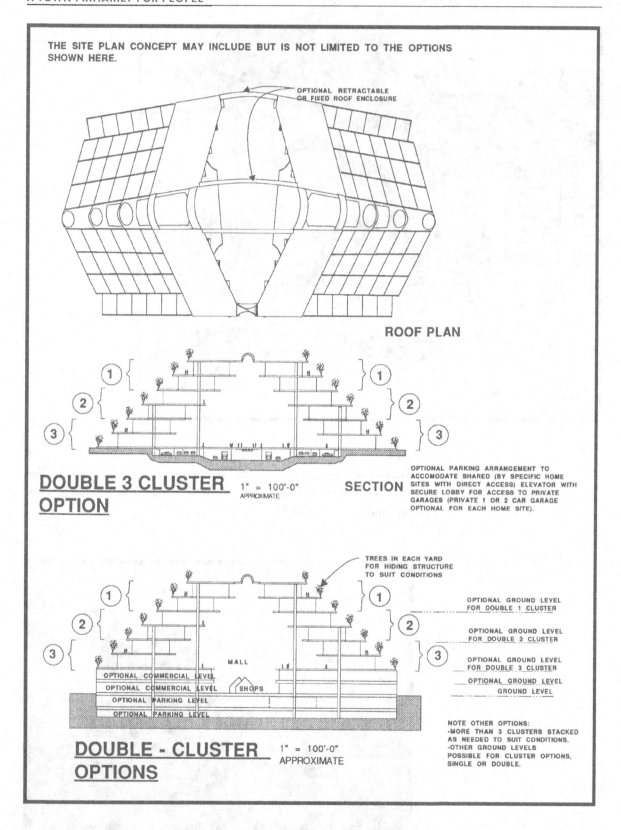

OPTIONAL RETRACTABLE OR FIXED ROOF ENCLOSURE

**ROOF PLAN**

## DOUBLE 3 CLUSTER OPTION

1" = 100'-0"
APPROXIMATE

**SECTION**

OPTIONAL PARKING ARRANGEMENT TO ACCOMODATE SHARED (BY SPECIFIC HOME SITES WITH DIRECT ACCESS) ELEVATOR WITH SECURE LOBBY FOR ACCESS TO PRIVATE GARAGES (PRIVATE 1 OR 2 CAR GARAGE OPTIONAL FOR EACH HOME SITE).

TREES IN EACH YARD FOR HIDING STRUCTURE TO SUIT CONDITIONS

OPTIONAL GROUND LEVEL FOR DOUBLE 1 CLUSTER

OPTIONAL GROUND LEVEL FOR DOUBLE 2 CLUSTER

OPTIONAL GROUND LEVEL FOR DOUBLE 3 CLUSTER

OPTIONAL GROUND LEVEL GROUND LEVEL

MALL

SHOPS

OPTIONAL COMMERCIAL LEVEL
OPTIONAL COMMERCIAL LEVEL
OPTIONAL PARKING LEVEL
OPTIONAL PARKING LEVEL

NOTE OTHER OPTIONS:
-MORE THAN 3 CLUSTERS STACKED AS NEEDED TO SUIT CONDITIONS.
-OTHER GROUND LEVELS POSSIBLE FOR CLUSTER OPTIONS, SINGLE OR DOUBLE.

## DOUBLE - CLUSTER OPTIONS

1" = 100'-0"
APPROXIMATE

That will also be true of a Nana Node's play area. Therefore, Nana Nodes can be supported by a structure and in relation to each home still function as their ground level. Arches extended off of columns and with overhanging plants give the impression of tree trunks with sheltering branches. Every home has its front rooms and its porch overlooking the play area. The front porch and sidewalk are on a terrace overlooking the street on a lower terrace. This is very similar to hill towns in Spain, where an entire town can be on a series of terraces.

To achieve something similar to hill town terraces, three Nana Nodes are stacked and stair-stepped. If a second group of three Nana Nodes similarly stacked is placed with front porches facing the first group, a space is formed between them. This creates a space that is larger at the center. If you cup your hands, with palms facing each other with fingers about an inch and a half a part, a similar space is there, inside. The fingers resemble the stair-stepped backyards.

The section diagrams at left show a 20-foot horizontal offset at each stair step. Later studies have found for certain conditions a 10- to 14-foot horizontal offset works better. This can vary, depending on whether it's located in a hot or cold climate.

This partially enclosed space conveys a wonderful feeling, sheltering but not confining, similar to Betatakin, with a view to the sky from each porch, but in this case the sides are open ended. It truly feels like a special place, a somewhere, … like it would be great to be inside of it. With front porches, it's obvious it will function as an extended neighborhood. It contains a section of the street, which becomes its own local street with friends who will know each other, possibly for generations. But it welcomes newcomers, visitors, and shoppers to enjoy this neighborhood's one-of-a-kind place, … its own unique portion of Main Street.

This extended neighborhood arrangement with its section of Main Street will enrich the lives of every generation. There will always be something to watch; sitting on front porches will be continually interesting. Watchful eyes of caring neighbors make it a safe place for young people, for seniors, … for all ages.

The other side of the street is far enough away for people to feel comfortable looking across the way at others walking along or on their porches, but close enough to recognize friends and wave. There will always be people coming and going, greeting people on porches, pausing in the play area, looking around as they ride the play-area public-access elevator to Main Street.

Gradually everyone will begin to appreciate the value of cooperating to make their community a great place to live.

When repeated to make an entire town, the synergy of this extended neighborhood concept generates new possibilities. It is very different. For some it may take being in it to appreciate it. The multiple levels of activity, long and short vistas, overall variety in shape of the spaces, openness to the sky as desired, and periodic views off to each side of the countryside with hills beyond will make this a truly remarkable environment. It will be more comprehensive, in all that it offers, than any other choices now available.

135

# Main Street

*A stage for community life, … close enough*

The goal is to develop a street that serves everyone, encourages community friendships along the way, and becomes part of each extended neighborhood. Provide enough other outstanding features to attract buyers from all other possible street design choices.

### These Are the Basic Design Requirements:

- connect to everything
- attract participants of all ages
- help make the feeling of a home town a reality
- have comfortable and wide enough surfaces for walking
- allow freedom for shop designs and changes over time
- provide vantage points for visual participants
- no cars

Primarily, a street is about a journey. It connects places. If it's fun, people will enjoy the journey. Then it can become a place, a long continuous place of movement, … similar to music, with patterns, loud, soft, high, low, … etc.

The street is made great by who, what, when, and how much it connects, including those who simply watch. In a small town, to be all it can be, it needs to connect everyone to everything and, in the process, to each other.

Design it so extended neighborhoods adopt a section of it. Many of these sections can become their own special place, ... a string of places along the way, something for everyone. Wider areas are for grand plazas, open-air theatre, great fountains, everything, ... even boats.

There should be a variety of ways to get from one end of the town to another. But there should be only one street, the Main Street ... the most convenient way to everything.

## How It Can Be Done

Imagining the experience will help in visualizing the possibilities of this entirely new concept for a street.

Watching the street from above can be almost as exciting as walking in the midst of it. Everyone in homes, on walks, or lounging on porches becomes a street participant, especially on celebration days.

But the street itself, wending through the various neighborhoods, gives each person an understanding of the character of each area of the town. In this setting, greetings are spontaneous, reinforcing the feeling of being part of the community. Much wider areas of the street are available for the public with concert shells, performance stages, boat rides, art exhibits, large gathering plazas, and people.

Every home overlooks this street; this is truly Main Street. At times it will very probably serve the purpose of the great parading avenues of Paris and London, where people crowd the streets to see and be seen. It could replace dragging the main in the 1950s; but here, people will be walking rather than inside cars.

Most journeys will start from the individual home site, moving out along the terrace sidewalk past porches, greeting neighbors, watching friends on the street below, planning a stop to say hello, passing through the play-yard, nodding and smiling greetings, going down in the open elevator, calling to a friend on the street below to wait, looking up at neighbors on porches across the way, moving to the center of the grand space of your own neighborhood, smelling coffee and fresh bread, brief visits at the café and then on toward your destination.

Having one street has one drawback: some things will be in one direction and some in the other. So there are choices for the return, perhaps the jogging trail along a hilltop ridge, or shortcuts along streams through the woods. But chances are, the street will be the chosen way to return because it's a continually changing experience.

This Main Street is a living place that will grow the community for the future. Shops can be freely refined over time to reinforce vitality.

It's the kind of place were children can go anywhere freely with their own means of transportation: *feet*. It provides a setting where they can have independence under the watchful eye of neighbors they know and who will enjoy watching them grow up. It is also great for seniors.

The space of each extended neighborhood is about the same size as inside the largest cathedrals in Europe, but much wider in the center and open to the sky. This space is only partially noticed from the street that passes through it. Street trees and shops with apartments above screen most views of the home terraces.

The main street level has home sites on both sides of the space for Main Street. The street area is like open land to be laid out for shops however desired. Buildings could be any design with apartments or offices above, up to a maximum of three stories. Some shops will be connected to the first level home sites. There will be one main circulation pattern, but for variety and interest walkways could occasionally jog off behind front shops and run parallel.

The opposite sketch is viewed from 30 feet above street level; the home site structures on each side with their walkways, plants, and porches are obvious. The view below, looking ahead down the street, is at the normal viewing level of 6 feet, so the home site structures are barely visible. Streets can be designed to seem like almost any typical old world village or the most sophisticated urban avenue. To see front porches requires looking perpendicular to the street from certain locations.

Approximately every 300 feet there are grand open spaces extending off to each side of Main Street. The views from the street and each

level above are only interrupted by two terrace-like bridges at 40 and 80 feet high (the play areas) and possibly a limited section of roof structure at 120 feet above street level. Those vistas are about 50 feet wide. This is at the widest point of the street and largest opening to the sky. This area allows streams of sunlight to enter Main Street most days and seasons.

Between those larger openings off to each side are narrower vertical spaces for access ramps from street level to the roof. These connect to walkways in front of homes at each play yard level and to some backyards. From landings along the ramps are great vistas of Main Street and the countryside.

These unique combinations of features will attract the interest and efforts of many people and give this street the potential to grow into a great street.

*Part Five*

# Habitat
# Support Systems

# More On This New Concept Town

*Space, shapes, and potential*

The previous essays talk about the habitat spaces and their relationship to the inhabitants. What it might feel like to be in it should begin to seem a little more understandable.

The diagrams show the basic architectural arrangement of spaces. The sketches and study models (on the facing page) begin to give some suggestion of the unlimited design variations that are possible. These are not intended to suggest the final architectural design character. Hopefully, as this concept's unique and various arrangements of architectural spaces becomes more fully understood for its revolutionary potential and practicality, others will be interested in working together to develop their own particular architectural statements. It offers the entirely new paradigm human habitats need to be all they can be.

These next essays will outline the integration and comprehensive nature of the elements that make everything function properly. Then, the next group of essays describes how the overall town concept will fit in several different locations.

If you still have a set of fixed stereotypes of what is acceptable and what is not, limited to things from the past, you may be wondering "what on earth is this?" Because, at first, it's going to appear as something it's not.

For example, some probably think it's just another "mega-structure" and believe anything of that scale is automatically bad for housing. Certainly past experience with incompetent design has left that impression. But this concept is different.

Even though this concept is multi-story, every people-occupied area has an intimate human scale that feels comfortably connected to nature and usable exterior spaces. Each home is like being in a two-story house on a hillside terrace. The neighborhood play area and backyard of each home are visually, psychologically, and functionally its ground level.

Living in this new concept town will provide the privacy of a home in the country, the convenience of a home over-looking Main Street, and the excitement of being part of a very livable town.

SECTIONS                    PLANS

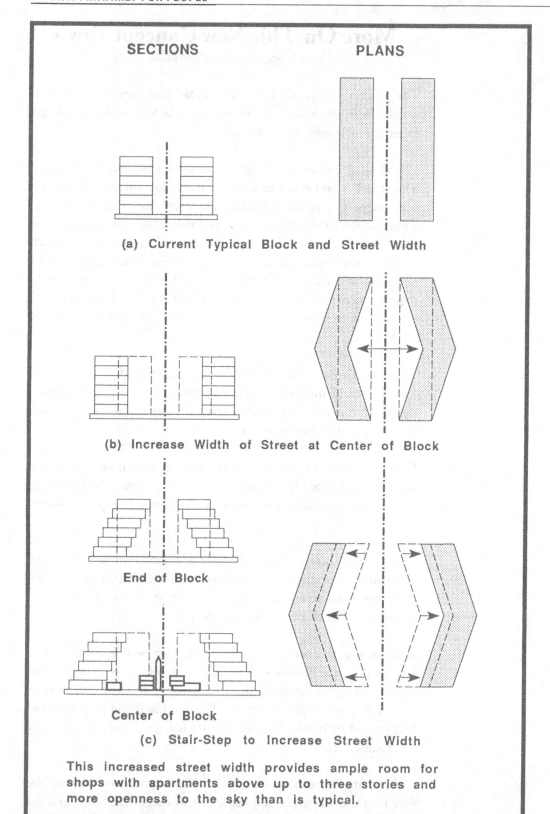

(a) Current Typical Block and Street Width

(b) Increase Width of Street at Center of Block

End of Block

Center of Block

(c) Stair-Step to Increase Street Width

This increased street width provides ample room for shops with apartments above up to three stories and more openness to the sky than is typical.

To generate totally new potential for livability, the first step is to get rid of the in-town car by making it entirely unnecessary. Next, get rid of all those wasted secondary streets by stacking four to six blocks of commercial-support housing above the one commercial block. Connect that to similar blocks to make a town.

That single combined street then becomes the most convenient pathway to everything: schools, offices, services, community centers, recreation, etc. It has a great concentration of potential shopper foot traffic, ... it has it all. This generates commercial success, which attracts quality shops that will be able to provide everything for everyone within a short walk.

But the greatest beneficiaries of these arrangements are the individual homes, each of their inhabitants, and their extended neighborhood.

Once again imagine the shape of the extended neighborhood.

Starting with one block of a typical street and multi-story housing, (a); at the center of the block widen the street to double or triple the normal width, (b). The ends of the block remain close together; the middle is further apart, creating a diamond-shaped open space in plan view. Next, starting at the top, step back each lower level of housing so that at street level the width is much greater, (c). Street-level housing is separated by four to five times the normal street width. This stepped-back arrangement gives every level an increased uninterrupted view of the street and creates a grand feeling of space. Add front porches to every home on every level, with each having a view of every other porch and the continual variety of activity in this grand space. Along the street separate commercial buildings are built; some shops have apartments above or are connected to the first level of housing.

This concept has the potential of being the most successful street every imagined for shopping, enjoying, and watching. Each block would be a truly grand extended neighborhood; the street is its living room, ... all day every day. Perhaps it's a higher-quality front porch version of the classic movie "Rear Window." In any case, ... definitely not boring.

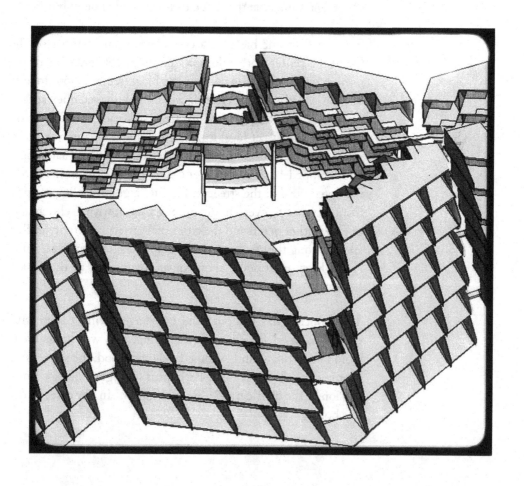

# Permanent Infrastructure

*Costs less than any other concept, for many centuries*

The goal is to develop an infrastructure that has minimal or even no recurring costs. Provide a comprehensive format for all utility systems so maintenance and upgrading is easy. Provide spaces in a planned design or armature that allows economical construction and easy remodeling of buildings for every livability function of a town. Provide enough outstanding features to attract buyers, investors, designers, and builders away from any other subdivision choices.

### These Are the Basic Design Requirements:

- design an infrastructure that is protected from the weather
- make it permanent — today this is physically possible.
- arrange it so it's easy to maintain
- provide space for every imaginable or future utility, recycling system, and service technique.
- make spaces flexible and interchangeable for homes, work, service, and all town activities.
- design a functional concept that will allow the most energy-efficient town ever built for now and the future.
- provide all this at a cost less than that for any comparable current town with related housing and facilities.

Design it so at least 70% of the initial investment in an individual home is not diminished over time, ... even for several hundred years.

### How It Can Be Done:

If we can do this, it could be the most revolutionary breakthrough for human habitats ever imagined.

Affordability will become economically and socially practical.

In our typical town subdivisions, there is enough money spent on materials and maintenance for all the streets, curbs, gutters, sewers, water mains, wires, cables, and sidewalks, plus the driveways, front yards, patios, fences, foundations, floor slabs, sidewalls and roofs of each house, to build the infrastructure for this new concept town.

149

The infrastructure consists of every element imaginable that serves a long-term need and will not need to be changed. Everything else is free to be whatever it needs to be, … any time.

Cost savings makes the permanent infrastructure approach more favorable than any other approach for a town. Building for the long term justifies a better quality of construction and design. The initial construction cost may be a little higher, but that can be offset by savings in financing and maintenance. Because of the long-term permanent value, more favorable community-based bond-type financing will offset a significant portion of that premium cost. Initial individual home costs will be similar to a comparable suburban home.

Over time, *continuing use* of the primary structure is economically much more efficient than any other alternative, including any attempts at reuse or recycling. There are no removal, disassembly, transport, or re-manufacturing process costs. Once that initial bond is paid off, that portion of the costs is free for the next several hundred years. And if it is 70% to 75% of the cost of each home, that will be an incredible savings. It could be the only way to catch up with our future growing and replacement housing needs, … affordably.

Imagine eliminating the current cost of maintaining and replacing all those typical items mentioned above. The savings may help pay for the local government and services.

Additionally, there are other savings to the individual homeowners. They build only what is necessary to meet needs and budget. When no longer needed, there's the potential for rental income. There's also the potential for sweat equity and investment for increasing family security and assets. Having only one-fifth of the home exposed to the weather creates a significant savings in energy costs. The savings from convenience and enhanced social value is also greater than can be measured.

The needs of a home site control the basic division of space. It will be a 20' high, two-story loft space, about 30' wide and 70' deep, with the outside end open to the countryside. It is expected that over time these spaces could serve many purposes, anything from hotels to light industry. Wall, floor, and utility systems are changed almost as easily as moving furniture and plugging in a lamp. Easy access to utilities and services is also a factor. Connecting openings in portions of sidewalls can be made as necessary.

The structure itself is fully protected by the enclosed spaces. Being fully protected, it requires no maintenance. The front or street sides of homes are protected by the overhanging stepped structure of the home sites above. This means the fronts of the homes will also be more economically maintained.

The structure may look similar to a honeycomb, but with square-cornered spaces rather than hexagonal. With the stair-stepped arrangement, the weather extremes affect only the back wall of each home site and a small roof portion, which is the backyard above it. The patio of the yard above will have translucent solar-collecting floor tiles, generating electricity and allowing light into the center of the home below.

The structure will be built of concrete, steel, and new lightweight epoxy cements reinforced with new high-tensile fibers. It will use connected components and on-site continuous-pour forming systems. Continuous-pour systems reduce connection cost.

On-going research using pneumatically supported forms have been the subject of ongoing research and testing, resulting in a 35% reduction in the weight and cost of this unusually efficient and very unique structural system. It would be very appropriate for building the infrastructure.

The long-term and systems nature of this concept make possible and has already spawned several other potentially new industrial ideas that will reduce initial and remodeling costs for owners.

The lowest level is half below original grade. The removed soil is placed around all edges, providing minimal dirt-moving cost and protection from flooding if necessary. This level is for service, storage, private garages, and light industrial. Any space could be used in the future for other activities. Service vehicles will have a tall space under the main street and plaza areas, which form the center service spline of the town.

A horizontal utility chase in a 15' wide continuous circulation pattern would be located along both sides of the service drives. This could hang seven feet below the main street level. It will contain all power, water, waste, recycling, cable, pneumatic delivery systems, etc. They are protected from deteriorating conditions of soil or weather. Utilities are easily inspected, maintained, and upgraded. The utility chase connects to all vertical shafts that connect to each unit of space above. Below this, continuous along the outside edge on each side of the central service spline, the lowest level extends into one- or two-level parking, storage, or workspace. Private garages are on the lowest level with elevator access to each home site above.

This town's infrastructure demonstrates how a three-dimensional division of space increases opportunities for optimal solutions and reduced costs. But like many other elements in this concept, its ramifications are multifaceted in application and value. Altogether, in ways not previously possible, it allows every element of a town to be more livable, practical, and efficient.

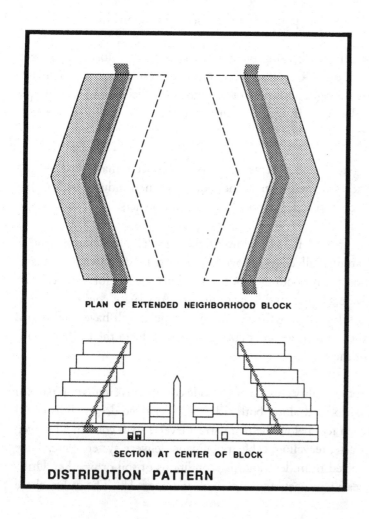

PLAN OF EXTENDED NEIGHBORHOOD BLOCK

SECTION AT CENTER OF BLOCK

**DISTRIBUTION PATTERN**

# For All Environmental Issues

The goal is to develop techniques for every environmental challenge so the town can fully integrate with nature. Provide an overall comprehensive and synergistic functionality where systems balance and reinforce each other as well as the natural surroundings. Provide conditions where systems can be easily maintained and updated. Make those features real, obvious, and understandable enough to attract buyers.

## These Are the Basic Design Requirements:

- provide summer sun protection
- use direct sun, earth, wind, and bio-energy
- convert and distribute solar energy for various uses
- recycle, reuse, or maintain everything used in the town
- use passive techniques to improve hot and cold conditions
- lay out all elements so maintenance is simple

## How It Can Be Done:

Meeting environmental challenges can never be successful with only a prophylactic or partial approach. We must fully develop an understanding of our responsibilities and potential.

Below are brief non-technical descriptions of some possibilities in this new concept town.

Its intended that this overall design concept will accommodate every objective and need of a town's environmental challenges. Efficiencies of every component are enhanced by the compact arrangement. Efficiencies are also increased by the open space and the shade trees in the back yards. Back-yard trees function like the town's exterior skin, providing shade in the summer and letting in winter sun when the leaves drop.

Compact, easily accessible utility passageways will make development and maintenance economical, as well as simplify the application of new systems in the future.

Utility passageways are designed for small-vehicle travel to facilitate maintenance and upgrading. They are large enough so new cables or tubes are easily installed without disturbing anyone. They connect to every building site.

Large pneumatic tubes connect most locations to each home. Computer-controlled capsules know what home cabinet to go into, … hot, cold, or general storage. This eliminates needing a vehicle for shopping. It adds convenience; a shopper can do more things on one trip and take time to visit with friends; no need to rush home with frozen foods.

This same system also carries away disposables. Capsules go to proper locations for recycling or reuse. Nothing is lost. Systems include every possible option for waste—from energy production, reusable materials, making new products, and soil additives to reusable chemicals and gases. Many materials have continuous patterns of use, recycling, and reuse directly within this town. Some may be sold as products for better use elsewhere.

The trees, essentially covering the entire structure, serve several purposes. The primary is providing seasonal energy-saving shade for the town. The second is providing the visual foil, making the town look like a tree-covered hillside from a distance. The third and bonus would be to provide a crop. This is consistent with the attempt for every element in this town to achieve an optimization and harmony similar to that found in nature.

Heavy waste water is piped to multiple ponds for natural processing with minimal energy or chemical treatment. The electrical potential between the waste water and the pond-bottom sediment is even used to generate electricity while aiding in the decomposition of the waste. After appropriate levels of treatment, water is reused for landscaping, farming, ponds, and even drinking.

Waste and mulches are used in sophisticated systems that will permanently maintain the surrounding farming soils at the highest level of production quantity and quality, … for hundreds of years. Water is used at least once before applied to growing food. Even then it is done in a manner that conserves water. Subsurface application under membranes provides the precise amounts of water needed and minimizes water loss. Greenhouses with dew-attracting membranes can recapture air-born moisture. The 45% translucent photovoltaic roofs

provide electricity, capture direct heat for related uses, and retain water. Being permanent makes this affordable.

This town uses up and loses a minimal amount of water. It just borrows it for awhile for various needs and uses it as a transportation vehicle to carry away related products that can be used in other places. After other useful applications, some will return to be used again.

Cleaner used water is treated at the home site and applied to the backyard growing surfaces. It's part of a system using the normally occurring evaporation to enhance interior home cooling. As mentioned before, translucent solar-collecting backyard patio tiles generate electricity and function as skylights for home below. Back walls can be solariums as transition spaces from interior to exterior temperatures and to warm or help cool the home.

Sunlight collected by sun-tracking mirrors on the top level is reflected to a central distribution pod on a roof of shops in each extended neighborhood. With computers it is then redirected to receptor locations at each home. Concentrated narrow beams of light go unnoticed except where they are needed. Direct sight-lines are made easier by the upper home sites overhanging the lower levels. Each home is able to redirect specific portions of that sun light to where they need it, even inside the home. It helps in keeping plants along the porches and inside the homes lush and healthy.

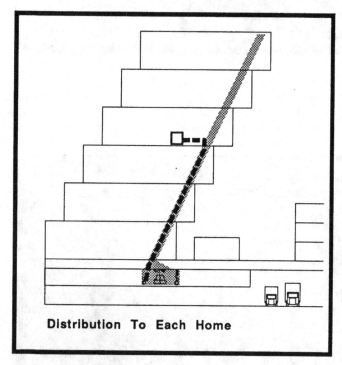

**Distribution To Each Home**

The main street can be partially enclosed as desired for various climates and seasons. Houses form the side enclosure. They provide an excellent layer of insulation. In certain climate conditions, the space at roof level between sections of houses will have lightweight, translucent, retractable structural fabric. Retractable structural fabric over the main street can contain heat in cold winters, and partially retracted can provide shade in summers.

In hot climates the street will be cooled like the deep palm canyons in the des-

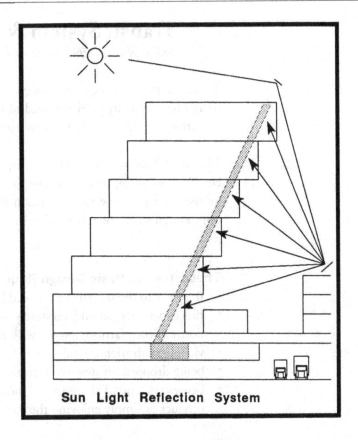

**Sun Light Reflection System**

ert. The night sky will draw radiant heat from the street level. The contained lower strata of cooler air, protected from disturbance by homes on the sides, will settle around the radiantly cooled surfaces. Gravity ventilation will carry upper-level heat out the top, and where appropriate can be replaced by evaporatively cooled air drawn through moist filters at the Nana Node pool areas.

The possibilities are unlimited and this new concept town will be ready for them. Current short-term solutions will not meet long-term challenges. This concept's long-term thinking makes multifaceted objectives and future combinations of refinements practical, affordable, and sustainable.

# Transit System & Connections
*Self-supported, efficient, no need to wait for regional rail*

There's a race between car design and town design; mass transit design has difficulty getting ahead of these. A town's layout is transportation's first and single most important design element.

The second element is the entire trip. To encourage transit use, the challenge is to improve upon the time spent and pleasure from the moment people leave their homes until they arrive at their destination, as well as on the return trip—no mess-ups and no waiting.

**These Are the Basic Design Requirements:**
- the walk to the transit stop should be a high point of each day,
- the experience should continue in the transit vehicle, settling down in your favorite lounge with the morning paper,
- visiting with friends, and
- being dropped off directly in front of your destination,
- being transported should seem incidental, almost a secondary by-product of simply enjoying the day.

The typical current suburban procedure may involve walking into your garage, backing out your car, putting on a seat belt, paying attention while driving, looking directly ahead like all drivers, waiting in traffic, recognizing a neighbor's car beyond waving range, often driving in heavy freeway traffic, getting cut off by another car, making a little smog, dodging through traffic, driving 15 to 60 minutes or more, finding a convenient parking place, and walking a short distance to the destination.

The ideal might instead be something like this: Leaving your home's front porch, greeting friends, pausing at the coffee shop, picking up a newspaper, a short enjoyable walk, settling in a lounge chair just as the transit leaves, more greetings, a bit of conversation or reading, 15 to 60 minutes or more passing hardly noticed, then exiting the transit directly at the front door of your destination. Returning home is basically the reverse of this.

## How It Can Be Done:

So how do we achieve such an ideal? Actually, a transit system's success is totally dependent on its convenience to each home. Everything else is secondary, including whether it's a car, bus, tandem buses, train, subway, or whether it's fancy, fast, slow, expensive, or cheap.

The layout of this town concept lets you get to the transit stop easily and quickly, and it provides opportunities to do this in a variety of enjoyable ways.

How do we handle the timing and get you directly to the front door of your destination? When you need to go anywhere, you tell the computer. It tracks travel times and destinations desired by everyone in the entire town; it groups people by destination and time, arranges the vehicle, and then tells you the exact time to depart. Alternatively, it could simply leave in several general directions every 15 minutes, and you tell it your exact destination when you get on.

The secret to making this work is the compact town layout. The transit boarding area is a single location; it's convenient to everyone and everyone leaving the town uses it. Walking there can be partially protected, so it is comfortable in any climate. The high frequency of use makes it practical, efficient, and affordable.

In this new town concept, having a single departure and return location makes it practical for the transit vehicle to make several front-door drop-off and pick-up locations at the other end. The computer, based on your instructions, would arrange the return pick-ups and notify you minutes before it arrives.

Each transit vehicle is selected according to the number of people going in the same general direction. As regular trip patterns develop, optimal arrangements will be made for each individual by the computer. Occasionally a travel group might develop. There will be a variety of vehicle sizes, from four to 15 passengers. Larger vehicles will be used for the most frequent destination. At farther destinations, vehicles may decide to park and wait to make the return trip.

This new concept town will operate its own transit system. This will contribute to its being more energy efficient and less polluting on a community scale from the beginning. The transit system is designed to accept transit conditions as they are, as well as to adapt and function with the inefficient layouts of other existing towns and cities. It cannot afford to be hampered by waiting until mass transit for the rest of its region is completed.

This town can be located in a new growth area where a transit system is planned but not immediately available. Initially as described, the system would use existing roads. To improve speed a lane could be added in the center of or adjacent to existing roads, to be used only by carpool vehicles. It could have travel in one direction in the mornings and the other in the afternoons and would have limited crossing intersections. These could be used by residents of other communities, if they traveled at the proper speeds.

When the regional transit is installed, if its stop is not directly connected to the town a short shuttle link would be available. It would be automatically more efficient in this town because of its convenient location near to each home. The major transit stop would probably have other regional commercial facilities planned near it.

In the future, if other destination places in nearby cities or car-oriented traditional towns change to have more compact arrangements, at some point in time a more-expensive regional mass transit system with it's own track/road/rail may become practical. With typical current densities in existing towns, or even in newer towns with layouts based on cars, a mass transit system will never reach maximum efficiency or be self-supporting. As a result, over many decades some entire areas of cities will probably decay and be available as future development sites for this new town concept.

This new concept town has the flexibility to convert to and make maximum use of any future transit system. It can work with the limits of any type of transit—train, bus, subway, mini-bus, private car, or even short-term rental cars. Any necessary changes can be allowed for and have minimal impact because any tracks or modifications will mostly occur in the surrounding open space areas. This is another example why this compact town arrangement with adjacent open space is so practical.

Older car-oriented sections in nearby regions will remain for a very long time, and so can this new town's original and very flexible vehicle transit system. The number of vehicles will adjust to the demand. As destinations change, this system can adapt.

But there is another possible challenge in this transportation discussion. Improvements in cars could delay the need for existing towns based on cars to change to mass transit. Non-polluting computer-driven cars that can automatically take kids to school and allow passengers to read on the way are possible today. Cars can change quickly to solve problems and meet people's needs long before a mass transit system can be built or a town can adapt to a transit system. Improved cars may help suburbia remain practical longer, particularly the New (Traditional) Urbanism versions of suburbia.

The bad news is that unless towns become adequately compact, once ideal cars are available, mass transit loses another argument for being the best future alternative.

The good news is that when ideal vehicles are available, their first and most-affordable use may be as the automated transit for this new concept town. Additionally, the surrounding region may be the area that benefits the most from the minimal traffic impact generated by this type of town.

This is another reason a town like this, on a site suitable to its needs, can be located in most places without having any negative traffic impact. It is a good neighbor anywhere. It even looks good. Actually, relative to traffic, positive impacts occur, whether it is located in large infill locations or in remote locations. It could even add greater use and value if located to share an existing in-city transit stop.

Or it could be miles away on the other side of a hill, with a single connecting transit element being its only access. Even then it could add benefits to the connected city. Its primary stop in an adjacent city would become a focal point of activity.

Villages like this new concept that do not allow cars to influence their layout and are designed to effectively use transit will help create and sustain points of interest.

Transportation will be effective because of the layout of this permanent new concept town. When there is a group of towns with a similarly convenient transit system, together they can support a new major civic and other regional centers, another reason for very-long-term planning. This will add a new dimension of quality to each town and an increasingly greater purpose for the transportation. This all works as a self-reinforcing, full-circle phenomenon.

# Integral Farming System

*Synergy of town and natural surroundings, sustainable*

A unique farming operation is needed to be an integral part of this town and its natural surroundings. The farming operation shall contain enough outstanding features to attract farmers from all other farming choices.

## These Are the Basic Design Requirements:

- function as part of nature
- supply food and resources to the town
- recycle by-products and waste from the town
- fulfill its role as a part of the town's sustainable systems
- provide employment, volunteer, and recreational opportunities
- add viewing interest to the open space

## How It Can Be Done:

This farming operation is about people, their town, and nature. Its daily operation will demonstrate simplified and understandable concepts for how people can work with nature. It can become a foundation for appreciating how the entire town works together with the seasons and the total environment. At various times of the year, there will be opportunities to work in planting, thinning, and harvesting. This offers young people seasonal employment. Just going for a walk down a country lane and observing the changing seasonal processes will be rewarding for all ages.

On the far edge of the surrounding open space, there could be a dairy and several complete working farms with all the common animals and crops. The farms could vary in size from 20 to 40 acres. They will be individually owned and managed. Some farmers may develop a specialty in certain crops best suited to the location, climate, or what the people find most desirable. Farm owners may live in town or next to their barns in farmhouses with rooms for guest visitors. For city dwellers the proximity, perhaps a 20-minute walk, makes available many experiences familiar only in small rural towns of a bygone era.

All the farming operations, totaling about 300 acres, will be a major element of several combined energy-saving and waste-processing systems. They will utilize waste products normally costly to dispose of. Each product will have its appropriate processing. Products will be purchased from the local processing companies, delivered in pipes or bulk.

Waste water will enter specialized ponds, utilizing natural processes to treat the water. There will be new, useful by-products. Microbes will treat the first stages. Water will flow through beds of plants and ponds with cultures of algae and other tiny life forms. At certain stages, the water will be suitable for raising fish and waterfowl in farm ponds. Eventually most of the remaining water makes its way, after appropriate processing, to a distribution system for watering crops or other reuse.

Some of the community waste will result in refined and beneficial products that can enhance the quality of the soil. It will be used on farm and recreation land. Soil which may not have been particularly good for farming can be turned into maximum-quality, high-producing farmland. This demonstrates the advantage of compact housing in close proximity to farmland. This is a great benefit for the land and its crops, ... for the town and its people.

This new town concept can replace suburban sprawl and its many disadvantages. From a long-term 500-year perspective, we will soon realize that some conventional suburban housing is basically temporary and too costly to maintain.

Once all that massive amount of suburban concrete and asphalt covering 40% to 50% of the land is removed, any old repairable water mains can be used by the farms for irrigation. For permanent maintenance of soils over hundreds of years, certain conditions will require drainage systems to avoid mineral buildup. Old sewer mains can be repaired and used for that purpose.

Housing and towns over time will need places to build. Reclaiming environmentally unfavorable suburbia could be the new frontier for future housing and towns. Suburbia's greatest value to the future may be the temporary nature of its construction. Much suburban housing of the twentieth century covered over some of our best-quality farmland. Actually, suburbia may be preserving good farmland for future generations. This new town concept leaves 70% of development land as open space and farming, making it possible to reverse mistakes of the past. Then, ironically, farmland may encroach upon suburbia.

This town concept gets its inspiration from the wonderful old-world experiences of farming villages and organic farming, plus the application of appropriate and advanced science. It creates an entirely new farm-to-town relationship never before possible. This will be one of the key ingredients needed to enhance its potential for a sustainable future, … fulfilling people's needs and real-world practicality.

# Other Features

*Maintenance, delivery, firefighting, energy, future systems*

The compact layout of this town allows for efficiencies and services never before imagined as being even remotely possible.

The infrastructure and utility passageways allow for all imaginable types of future systems. That could include home-to-home delivery, or even a fresh crop sent directly from the farm if it was too late for a nice walk before dinner.

Directly under the street-level shops will be their storage areas, workshops, or additional shopping. They may have two levels in the two-story space. Service vehicles' circulation patterns will be one way along one side of the storage areas with the return direction along the other side. The circulation will loop all the way to the end of each branch of the town. Cross-connection between the one-way patterns may occur wherever necessary.

There will be unloading areas for trucks next to the storage. Passenger vehicles will use the same one-way pattern. The clusters of parking will branch off to each side. The parking will be located directly under the respective residences in private garages as desired. There will be several options to suit needs at the time. Residential parking can be on the upper of the two possible levels or the lower. Parking may be stacked with private lifts within the garages, requiring less floor area. One of the levels could be used for light industrial; in some situations this could take an entire floor level. Essentially all of this is flexible space, easily changed to accommodate changing needs over the long term. This also provides for the possibility that the nature, use, and size of cars may change significantly.

There will be a two-foot or larger continuous horizontal space for utilities under the shops. The main linear utility passageways will run generally parallel below the residences on both sides of the main service vehicle corridors. These passages will be located to collect all the vertical utility drops that connect to each residence just below Main Street level.

The elevators connecting to each residence will be shared by a residence on each side and on each level of housing. They will require a key for access and each trip should be able to be private. Studies have shown that overlapping demand will occur very seldom with this number of residences. A sloping elevator shaft will probably be the best for the layout of each home site. New elevator designs have been recently developed. The utility runs will be located on one side of the elevator shaft and in its corners. The elevator will provide access to those utility runs.

Elevators are an expense that may not be necessary for all homes. The space for it will be there as a utility chase and for future use as needed.

Many residents will not need a private garage or car because all regular needs, including transit, are within an easy walk. Renting will satisfy the occasional need of a car. However, each Nana Node has an elevator to street and garage levels. This provides the opportunity to have a car during the initial transitional stage for new residences. In this case, cars would be in assigned parking spaces.

Along the top level adjacent to the ridge-top jogging trail will be a track for the portable hoist used to deliver large loads to each back yard. There will be loading terminals on the vehicle service level between pairs of extended neighborhoods for arrivals of large but lightweight shipping containers. These can be picked up by the roof top hoist and lowered into the back yards of each home. This would occur only during initial construction or major remodeling and for the delivery of furniture during moving if necessary. For normal moving, the elevators would be adequate.

Firefighting or other special needs may find value in these hoists. Firefighting would be rarely needed, however, because all construction will be fire-protected or fire-resistant. Even the back-yard trees are close enough to be protected by the fire sprinkler system.

The hoist is also used for maintaining the back-yard trees when necessary. If the two primary trees in each yard are fruit- or nut-bearing, they could each be lifted to the roof during the harvest season. The crop removal would occur on the roof, the tree would be replaced, and the next tree lifted up. The roof would also have a large percentage of the area that would be used for trimming, storage, and treatment or recovery of trees that required occasional special care.

Because the back-yard trees provide a near-complete vegetative coverage over every part of the town, 85% of the land is still used by nature. The plants at each front porch and along the residential terraces and Main Street add back perhaps an additional 5%. This human habitat removes vegetation from less than 15% of the land. This is remarkable if you think of all the land in typical towns used just for streets and roofs of houses; it certainly more than 50%.

This town layout adds or restores traditional uses to the home. For example, the home can become the best place for medical care at most stages, from birth to the final days. The compact design and advanced monitoring systems make this possible. Distances are close enough for a regular nurse or even a doctor's visit. Family care can reduce costs.

This town can be thought of as stationary cruise ship. It will have great views from every suite, plus more room and more activities. This town could become one of the world's great destination resorts. The townspeople would certainly want to place limits on that, but tourism could add to everyone's employment and enjoyment. This town will provide unique opportunities for new creative solutions.

Energy distribution for lighting may find other options. Some of the most efficient light sources can't easily be located near people. If a large, efficient source was mounted on the roof, however, at night it could use the same mirror system used by the sun in the day-time.

Initially the entire town could be shaped to concentrate wind currents for wind-power generators. They could be located along the top ridge of the town, or one end branch of the town could open up to accept a handsomely designed multi-story fan.

The need for easily changed wall and floor components offers the basis for an entirely new industry of manufactured devices. Some could be manufactured in this town. Of course there will be typical panels, but some walls could be like furniture or storage units that the homeowner puts together. They could have easily sealed connections to each other and include integral pre-hung passage doors. They could support floor panels. Bathrooms could be modular. A used one could be sold to a young couple starting out when someone didn't need it. It could be simply moved between homes by the crane.

This just scratches the surface of the possibilities. The most remarkable will be the ones we can't yet imagine.

# Existing General Plans

*Adaptable, with trade-offs for improved livability.*

Current General Plans can be easily modified to include this new concept town. Procedures and trade-offs could be similar to those used when Planned Unit Developments or condominiums are proposed in areas zoned for traditional subdivisions. Typically, an area of appropriate size would contain the same various residential densities and professional and commercial zones as shown on the existing General Plans.

These same uses would be provided in this new neighborhood community concept, but laid out with its comprehensive systems and new arrangements. The most significant trade-off: provide surrounding open space in exchange for allowing an increase in the normal height limits. The sight-line angle of the height above the horizon would be the same as the underlying zone and the total density could be the same.

Because this new concept actually makes density desirable, and with the scale of the open space, in certain locations it may be possible to increase the overall density or to include professional office space, artist studios, or even light industrial that may not have been in the original general plan. This would be a reasonable trade-off because more residents could then have their work in walking distance from their homes, reducing going-to-work traffic and the impact on the existing surrounding area.

If it became desirable, a hundred years or more in the future, plans could even allow high-rise towers. Because of the overall design concept, this could be done without any negative effect on this new neighborhood-type community or the surrounding area.

If built as infill in existing older areas and if the initial available area is not big enough for a complete town, an overlay specific plan would be done for the ultimate maximum area to be developed over many decades or centuries. The concept allows for gradually and efficiently phasing in the old with the new. Improvements to older but still useful facilities that are planned to be removed in the future would use modular bath or kitchen units and systems that would ultimately be reused in future new town home sites. That way, old facilities could complete their useful economic life with high-quality improvements, but the investment in such improvements would not be wasted.

Long-term planning makes higher-quality investments cost-effective and practical.

This eliminates the typical cycles of gradual deterioration into poor-quality neighborhoods until they are so bad the only choice is removal and replacement. Other trade-offs may also evolve to allow a different mix of uses and density. With truly long-term objectives clearly in mind, however, such refinements can be made with confidence.

Any initial development would need to be large enough to adequately establish a basic functional town and the sense of community.

Because of the special nature of this new concept for a town, existing General Plans may also be modified to allow development in areas considered too remote or outside normal growth limits. These limits were probably established because of the negatives expected from current development and car-based concepts. Because this new concept eliminates those negatives, communities may find unexpected locations to be very acceptable for new development. These could be areas currently not optimal for farming, but when combined with the integral ecological systems of this town new farmland acreage could be added that would be improved to the highest quality and be sustainable for centuries.

Another basic feature of this design concept for a remote area is that, at its own expense, it could financially afford to provide its own transit and integrated waste/farm system. Transit would include installation (as described in the transit essay) of travel-ways connecting it to major destinations. This could be a General Plan requirement.

It may also be acceptable to establish restrictions on the number of vehicles leaving the town. This would be reasonable, because a greater amount of work could be expected to be in town, within walking distance from each home. That's in addition to the existing trend toward working at home.

One of the main features of this concept is the convenient walking distance from each home to everything, particularly the transit stop. Any future regional transit could easily be integrated into the town and replace or complement the very flexible initial system.

Many communities are struggling with how best to control growth in the rural areas. Some have proposed one primary residence, a couple

of other houses, and typical farm structures on 40 acres. Over the next century, descendants and those who wanted or could afford that rural life style will probably scatter housing across the farmland as far as we can see. If we then tried to assemble a square mile of land for a town development, it would be very difficult and expensive, if not impossible. *This new concept town, with the same density as suburbia, would take less land out of production than those few houses and the required roads.*

States like California are already recognizing the necessity of planning for future growth. For many communities, the growth for which they are expected to provide seems overwhelming. But even the state's projections for required planning objectives are too shortsighted. It is almost as if the really long-term future doesn't exist. If we don't plan for it, perhaps it won't exist, … at least not in ways we might prefer.

General Plans that are trapped by current concepts and still based primarily on cars could be the biggest obstacle to building communities built primarily for people.

This new concept town, better than any other current alternative, offers a strategy to satisfy the often-conflicting interests of environmentalists, landowners, and developers, as well as the need for affordable housing. Permanence and sustainability give it economic practicality.

Sound General Plans can be part of the support system for natural and human habitats. This totally new concept town has the potential to meet our housing, social, and community needs, … now, in the immediate future, and for several centuries.

*Part Six*

# Habitat Sites

## CONVENTIONAL SUBURBAN SUBDIVISION (DIAGRAMTIC)

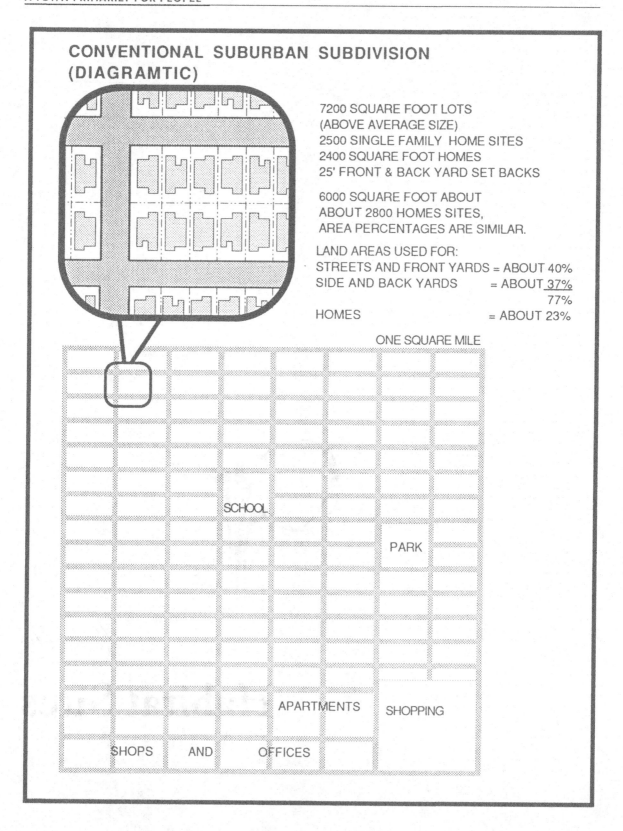

7200 SQUARE FOOT LOTS
(ABOVE AVERAGE SIZE)
2500 SINGLE FAMILY HOME SITES
2400 SQUARE FOOT HOMES
25' FRONT & BACK YARD SET BACKS

6000 SQUARE FOOT ABOUT
ABOUT 2800 HOMES SITES,
AREA PERCENTAGES ARE SIMILAR.

LAND AREAS USED FOR:
STREETS AND FRONT YARDS = ABOUT 40%
SIDE AND BACK YARDS      = ABOUT 37%
                           77%
HOMES                    = ABOUT 23%

ONE SQUARE MILE

SCHOOL

PARK

APARTMENTS     SHOPPING

SHOPS     AND     OFFICES

174

# Comparing How Land Is Used

*Removes less than 15% of the natural biosphere*

Perhaps our most important responsibility is how we use the land.

In this presentation, a square mile (640 acres) is used as the site for the prototype of this town concept. The purpose here is simply to describe and compare advantages of the concept. Except for being a familiar dimension, which may help in understanding comparisons, there is nothing particularly significant about this size. There are many other factors that would influence the size of a specific village or town.

United States suburbia is used for market comparisons in terms of house size, basic amenities, total number of houses, and land allocations. A square mile typically has about 2500 homes, some shopping, office space, apartments, and an elementary school. These are enough elements to make a small town or village. The objective is to rearrange everything in a manner that creates a wonderfully livable village atmosphere. It's primarily for people, no cars, … with a short walk to everything.

For other countries, it's understood the sizes and functional aspects of this new concept would be adjusted to meet local needs. The fundamental arrangement of spaces and opportunities for use would still be valid.

One basic problem with suburbia is approximately 40% to 50% of the land is used for streets and sidewalks. Including front yards, the area can often be over 50%.

Imagine if we could simply get rid of all of the automobile circulation spaces. Suddenly we have 50% of the square mile (320 acres) as open space. Usually that is already public land.

Side yards are basically for ventilation and privacy, but walls with windows next to other homes are often only five to ten feet apart. Building a solid wall between homes would offer better privacy and make another 5% of the land available for a better use.

Suburbia has a lot of open space. Unfortunately it's in long narrow strips. There is no way it can be a natural open space. How can this land, over 400 acres, be pulled together as a more useful space and still have the same number of homes?

Where's the best place for the street? Hide it. As a way to start, thinking abstractly, one wild idea would be to put the street and cars under the houses. Next, stack houses from one side of the street on top of the other side with street and parking below. But that makes a long, narrow circulation space with costly below-grade retaining walls.

However, if six houses were similarly stacked they would all share the cost of the street. The street could have private garages for each house, reached by a shared elevator for each pair of houses on each of the six levels. Then the expense for the retaining wall, street, and garage for each home is affordable.

This descriptive way of visualizing an entirely new arrangement between homes and cars shows how simple it is to save hundreds of acres of land normally used by streets. This, along with new arrangements for the way all the land is used, became one of the fundamental principles of this new concept town.

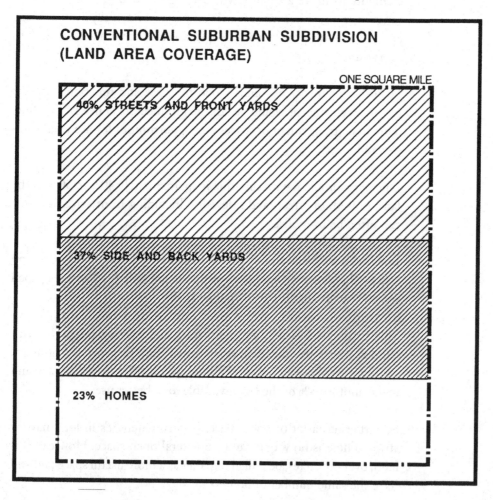

CONVENTIONAL SUBURBAN SUBDIVISION
(LAND AREA COVERAGE)

ONE SQUARE MILE

40% STREETS AND FRONT YARDS

37% SIDE AND BACK YARDS

23% HOMES

Over 70% of the land can be made available for open space and have the same number of homes as compared with typical suburbia that covers the land, chopped into little pieces between buildings and streets. This makes it possible to have space for farms, lakes, woods with meandering creeks, and outdoor sports. Most importantly, in addition to food production, this space is large enough to be an integral part of the water purification and recycling systems for the town. Such open spaces would pay their own way. Usually open spaces near homes in current typical developments are only for recreation and become a maintenance expense.

But for people who live in suburbia, stacked houses don't sound like a particularly desirable place to live.

So here's a brief review of possibilities. Let's give each house a back yard big enough for a patio, a couple of fruit trees, a garden, a swing set, and a hot tub. Then, rather than viewing neighboring tract houses

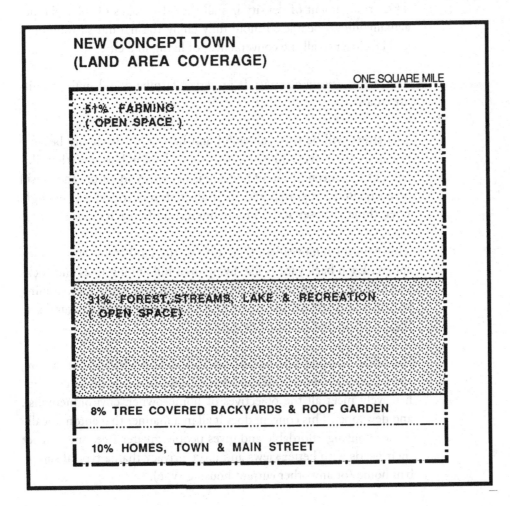

NEW CONCEPT TOWN
(LAND AREA COVERAGE)

ONE SQUARE MILE

51% FARMING
( OPEN SPACE )

31% FOREST, STREAMS, LAKE & RECREATION
( OPEN SPACE )

8% TREE COVERED BACKYARDS & ROOF GARDEN

10% HOMES, TOWN & MAIN STREET

overlooking the back fence, give each back yard a clear open view of several hundred acres of open space. It'll be like on a hillside in the country. Make it totally private with 20-foot-tall fences on each side sloped down to six feet at the back edge of the yard.

On the front of each house include a large porch, sidewalk, and neighborhood play area with all overlooking Main Street.

That main street will be wide enough to duplicate your favorite village street from anywhere in the world, or be a better one designed for the local culture. The street will be open to the sky.

When walking on the street, the taller stacked houses behind the shops on each side won't be apparent unless a special effort is made to look up and over between the shops. The street can be protected from the weather if necessary.

This arrangement of density has all the advantages of suburbia, but actually allows people to more fully enjoy the natural surroundings and be closer to all the conveniences of living in a city.

*This makes density desirable.* It preserves farmland and multiplies its usefulness.

The street will be very successful and have a lot of vitality because it connects to everything for everyone. Research shows that this amount of concentration of housing is necessary to ensure shops will be adequately supported and the transit stop is convenient enough. Everything is closer and it's more interesting for each home.

This concept removes less than 15% of the land from nature's biosphere. Each back yard is like living on a hill among trees and rocks overlooking the countryside. The front porches, with vines, arching branches and vertical forms, are also a lot like being in a natural setting—urban, but softened by nature.

No other housing alternative has this delicate balance of nature, spatial arrangements, and the level of density adequate to comprehensively meet all the objectives of home-owner, environmentalist, and developer. This concept does it all in a manner that automatically makes housing affordable and gives people greater freedom to meet their needs with better views and more privacy than a typical suburban home (or any other current housing type).

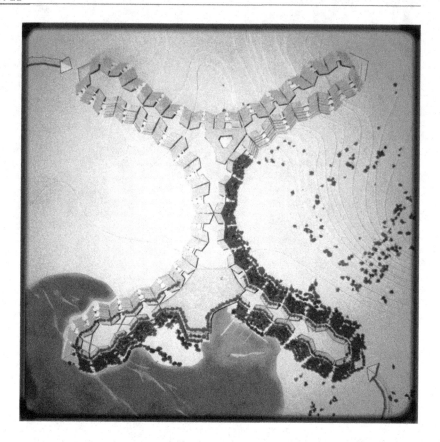

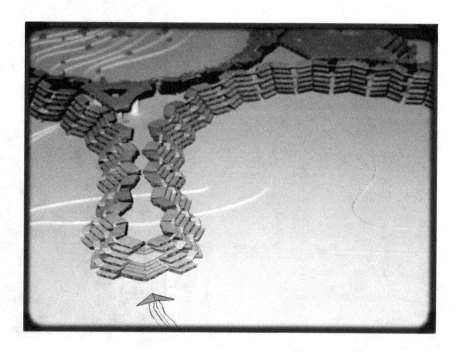

# Site Plan System

*Adjust for natural or man-made surroundings*

The goal is to develop a flexible site planning system that can adapt to site terrain features and become an integral part of natural or man-made surroundings—when viewed from above, from within the internal spaces; from far and near.

## These Are the Basic Design Requirements:

- look and function as part of nature
- gracefully accommodate all facilities needed in a town
- adjust angles between modules to fit site
- be a good neighbor to any existing communities
- accommodate future growth demands as necessary

## How It Can Be Done:

The town's master plan system, composed of Ext-n-Node modules, simply allows it to develop its own custom shape to fit the terrain of the chosen site. The angles at the central play area of each extended neighborhood module can be adjusted as necessary. The angle at the stair/ramp between neighborhood modules can also be adjusted. The maximum possible angles at each location have optimum limits to form an overall shape with a continuous flow, but any angle, or even separation, is possible.

At the planning stage, adjusting the plan of the town with its multiple connected modules is like flexing vertebrae to achieve the desired curvature. In an aerial view, especially without the exterior skin-like covering of trees, it looks like something that nature could have evolved. Once covered with trees, which function like its exterior skin, the angular modules are softened.

At intersections of three branches the modules are split along the street line, each half curving in opposite directions, until they join with another matching half of the adjacent intersecting branch. At these intersections the street is widened, allowing for grand plaza areas. These will become important areas of activity. They can accommodate a variety of community-scale functions or buildings. Any buildings over such main street locations would be on stilts to maintain the character of the street. The tall columns would get lateral support from diagonal struts integrated into the stair-stepped housing structure.

These wider street areas can also be located wherever appropriate along the way or at the end of a branch. For example, hotel rooms could occupy typical home sites, while the other required hotel spaces would extend into the wider street space. Then taller portions could extend to the desired height above the typical structure.

One enlarged street area will become the primary community business, service and education center. Similar to the description of the hotel, buildings can be within the primary structure or, in the future, extend above it.

The seven-acre elementary school playground is located on the roof over the main center. Playgrounds generate a lot of noise; placing it anywhere on a lower level would be noisy for someone. Here the angle of the sound would be shielded from almost everyone. The classrooms would be located just below, with windows overlooking the open space, lake or the main street plaza.

Over hundreds of years, future generations may not want to move from the town of their ancestors. This entire community can function as an exciting and thriving foundational platform for future high-rise

towers. It provides what towers need — to be integrally connected to this kind of Main Street vitality and to the other towers. Controlling sight lines can maintain the privacy of the backyards below.

This town with its back-yard trees and low silhouette, when viewed from beyond the surrounding areas of open space, simply blends into the landscape as if a natural hill. This makes it very acceptable as a neighbor in almost any location.

It could be in a rural area near other small towns and farms, in the middle of a vast suburban area, or ideally, next to any large city's downtown center. In suburban areas or on large downtown infill sites it will bring with it the qualities and benefits of a natural rural atmosphere. It will significantly improve the value of any existing facilities surrounding or adjacent to it.

This concept takes everything we've often hoped was possible to a higher level of quality and satisfaction. With the aid of more drawings, models, and video (perhaps a future DVD), the combinations and value of all the elements will gradually become more clearly understood.

Below is a one-mile section of the town on a flat site. On sloped or hilly sites, it can step up or be parallel to the countour of the site.

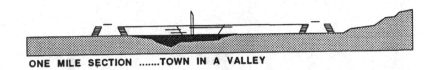

ONE MILE SECTION .......TOWN IN A VALLEY

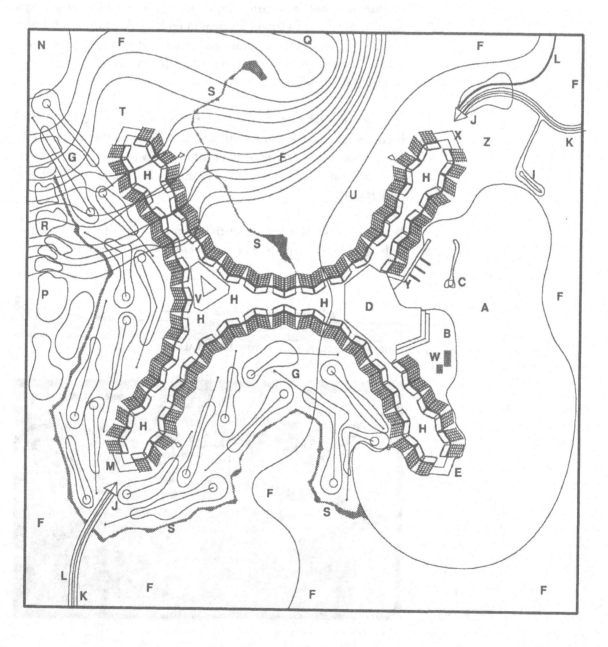

# Recreation Features
*Places to re-create spirits for everyone*

The goal is to develop optimum locations for all types of recreation and provide opportunities for people to enjoy and appreciate being part of nature. Hopefully they will come to accept as normal that humans are part of a comprehensive and complementing natural reality. Also provide space for sharing and competing in all types of sports.

## This Is the Basic Design Requirement:
- provide every imaginable opportunity for each person to participate in activities that uplift and re-create their spirits.

## How It Can Be Done:

This concept has unique places for all kinds of recreational experiences. The trail on the ridgetop of the town will attract many people. It will be less than two minutes from every home. Some will come just for the various views, beautiful sunsets, and stargazing. It will be like a top-of-the-world place to jog, stretch, or walk to other parts of town.

Another experience will be strolling along upper-level sidewalks, past porches, through extended neighborhoods while overlooking Main Street. It would be like being in some kind of a fabulous hanging garden. Walking past play areas reveals glimpses of the countryside through four-story arches, beyond openings overlooking lower levels, and the horizon is just above the sparkling water of the local swimming pool.

The interior of each home site has the flexibility and potential to fulfill the desires of its occupants for everything, including recreation. That will be enriched by backyard views of forest trees changing colors through the seasons, boats on a lake, or morning mist covering pastoral scenes of cows grazing in the distance. Similar views are possible from the play areas and the tall vertical stair/ramp passageways between extended neighborhoods.

The entire town is about experiencing moments that uplift your spirits. Looking out of front windows and stepping out onto your

**CONCEPT EXAMPLE**

**MASTER SITE PLAN**
One Mile Square Option
Example 1    640 acres
**LEGEND**
A - Lake
B - Beach
C - Clock Tower
D - School Yard, School
    & Community
      Facilities Below
E - Entertainment &
      Senior Center
F - Farm Land outside of
      tree line in
        Model, +/-320ac.
G - Golf Course
H - Highrise locations for
      Office or
        Residence Towers.
I - Vista Point
J - Roadway & Transit
      Entrance
K - Roadway
L - Line of Mass Transit
M - Golf Club
N - Utility Services
P - Ponds, Aguaculture
Q - Equestrian Center
R - Recycling Ponds
S - Creek
T - Tennis Center
U - Forest Park
V - Hotel, Special Facilities
W - Water Sports
X - Exercise Center
Z - Soccer Fields

porch, there's the warm feeling of friends nearby. It's the link to the neighborhood, an alcove of the community living space. The gentle transition of viewing, moving toward, and then entering the activities of Main Street becomes its own special form of recreation. That continues along the way to everything.

Defined recreational activities are a magnet for people. Some are noisy. The best place to locate various types is at the end of the town's branches. There they will be heard and seen from the fewest home sites. People will buy a home site near the activities of greatest interest to them. There will be clubs for tennis, beach and swimming, boating, fitness, soccer, golf, senior activities, and whatever people may want. There is plenty of open space. The Main Street plazas near each particular sport will have shops related to that sport.

Less than ten minutes from every home, there will be woods with overnight camping sites. These can be near favorite fishing or swimming holes. Walking trails and picnic spots can be along fishing creeks winding through the woods. There will be cottontail rabbits running across trails, deer, foxes, and a great variety of other native animals.

There will be orchards with the smell of spring blossoms and fresh fruit in the fall. The working farm and dairy will offer wonderful, uplifting experiences. Other activities like horse riding, rock climbing, and gardening round out recreational opportunities.

The entire L.A. basin could be covered by village towns like this, with at least 70% of the land as interconnected open space. With its magnificent climate, it could actually be better than before humans came, rather than worse as it is today. All this is possible for many places in the world.

This concept is about livability, ... every place, ... every moment, it's about lifting spirits. And that includes a lot more than just recreation.

# In-City In-Fill

*Desirable density with integral open space*

What would be the strategy for building this new concept town in or near the downtown center? The primary purpose would be to add housing and vitality. In many downtown centers this is badly needed.

This concept has a great advantage. The footprint of this new concept town with its housing is very small compared to the number of homes provided. If built in areas with old housing, it allows 70% of the existing housing areas to remain occupied until the new housing is completed. This is possible because its footprint is smaller as compared to any other current housing concepts (except, of course, for 60-story residential towns, which are for different objectives). The introduction of its surrounding open space will also enhance neighboring properties.

Another advantage is the potential flexibility of site arrangements. This allows designs that connect with existing historic structures or any city fabric that is desired to be retained. The town can be designed

to incorporate anything existing; some elements could be part of the new main street. Or it can be adjacent to an existing major feature or on a main pedestrian connection. A shared downtown transit stop can act as a connector. Certain streets or utilities may pass through this new residential structure if necessary, and bridges can connect to existing structures above street level. Any interconnecting circulation system between old and new, however, must be pedestrian.

This new concept town must bring with it all the features that will make it an ideal livable town. Initially, enough must be built and connected with appropriate existing facilities so together they will function as a complete unit with its own vitality. Generally, however, it will be best if the new site plan, patterns, and utilities are free of any existing streets. Each phase must be completed as quickly as possible. This type of development can restore enthusiasm and enrich every other element of a downtown area.

Most importantly, whenever we think about plans for the city center, we must force ourselves to think in terms of hundreds of years. There is no longer a "Wild West" without limits.

Urban designers and planners are trained to consider long-term issues. Those in the role of client must make long-term planning a requirement and budget for the additional time involved in that process. It requires a more comprehensive plan than is typical.

If the existing city center has any characteristics and features that are worth saving, those can become the starting point for future plans. Everything else must be considered and scheduled as disposable.

Many things have reduced the attractiveness of housing in our downtowns. The freeways bring people downtown to work but not to live. They make the suburbs more attractive. Reduced activity encourages general stores and business to seek other locations. Increased crime is often a fear. Offices related to or supporting government or court activities eventually move out because computers, the Internet, and freeways make remote locations possible. All of these are potential markets as buyers or tenants. Market surveys suggest many people would prefer the downtown urban life if there were a better combination of choices.

Trying to patch together relatively small-scale existing elements, or make incremental improvements in a piecemeal fashion, is not the answer to the nature of the problems most downtown areas face, particularly if they hope to have any long-term potential.

For this comprehensive new concept town a large area will be required. It does not need to be built all at the same time, but any phasing and the eventual overall design need to be part of the plan from the beginning. The initial phase needs to be large enough to generate enough activity to be functionally complete.

Large brown-field areas may be suitable for redevelopment. Areas with older service-type buildings that provided workplaces when the town was smaller might be ready to be abandoned; some probably already are.

Unfortunately, as these areas have become cheaper, new developments for senior or affordable housing often have been inserted in a spotty manner. Premature in-fill housing, today's pet idea, that is not part of a very long-term comprehensive development plan can also create spotty patterns of improvements. Several decades from now these may still be too new to remove, trapping the land and preserving existing or outdated patterns. This can delay or even prevent better comprehensive concepts from being built for a hundred years into the future.

Rather than let the available parcels control the design or location, thought should be given to what would be most desirable if the new pattern created was expected to be there for five hundred years. This kind of thinking might make it possible to work around location problems presented by scattered spots of development so they might be allowed to remain for their useful economic life.

Whatever is built must be clearly expected to be there for hundreds of years. This calls for everyone to be bold in their thinking but careful. This will give developers on nearby parcels the confidence to make the highest-quality investment, and also to be equally careful and thoughtful. Any open space approved should be included as an integral and permanent part of the development and the five-hundred-year plan.

# Enhancing Historic Places

*Historic preservation with a sustainable synergy*

Many historic small towns are in danger of being overrun, overcome, or torn down.

They vary in size, layout, and level of current community vitality. Some may have been locations of important historical events that need to be preserved for future generations. Certain buildings may be more important than others. In some, the entire setting of that past moment in history may be important to fully appreciating the time or the event.

The flexibility of this new concept town can make the preservation of a historically significant small town possible and economically practical.

There are many different ways the new town could connect to the old. It could be at one end of Main Street. The existing Main Street could extend into the new town as its Main Street. At a reasonably significant distance, the two new sides of the street could curve to open up wide views from the old town into the partially sheltered interior of the new town. The new homes' front porches would have a view of the old town in all the beauty of its old and natural setting. The front porches and elevated play areas of the new would be designed as beautiful hanging gardens, each home partially hidden by trees, vines and flowers. The top edge of the new housing could stair-step down so the nearest new buildings would be only two to four stories tall and fade behind the large trees on both sides of the street.

Alternatively, the new town could appear as a tree-covered hillside behind the old town. Perhaps its natural character would be a better backdrop, helping to maintain the historic quality of the old town by hiding the view of an adjacent city's high-rises just a few miles away.

The appearance of that tree-covered hillside can also be modified in ways that better complement the old town. A street from the old town could extend up the side of the tree lined terraces of the new town. There could be gradual slopes and steps as the pedestrian-way continued up the apparent hillside. Buildings designed in the character of the old town could be built along this winding, climbing pathway. It would appear as the extension of the old town street. At the top of the new town could be an important new community place. Next to it could be a small plaza with great views beyond the quiet setting of the old town and the surrounding farmland.

The path would offer a beautiful walk in the evenings. From all levels of the new town people could emerge from the modern urbane city-side porches of their homes and come out onto this gradually descending pathway, greeting friends and shopping along the way. As they stroll down the hill, the countryside roof-top view would change to views engulfed by the old Main Street. Their destination could be one of the historic restaurants, an old time dance or a town hall meeting.

Together these two towns from different periods have a future, helping to preserve the traditions of what will someday be an ancient town, ... eventually two ancient towns.

Often historic places are composed of a single structure that remains because it was better built or it was protected by an owner who appreciated its importance. The surrounding buildings may be of such poor quality that the single building is in danger of being torn down along with the others. This new concept town could incorporate a single building or a small cluster of buildings as an integral part of its new Main Street.

The new concept town can have very large plazas designed to accommodate many different historic situations. A historic building could be a focal point of the new town. It could be in a grand plaza facing the

center of town. Its backdrop could be the fronts of overlooking homes with their lush hanging gardens. Behind it the housing arrangement could have an opening in the form of a dramatic arch twice as tall as the historic structure with a view to the countryside. Rows of trees and paths could extend from the plaza on each side of this featured historic building out into the open space beyond. The rows of trees could continue along the promenade to a grand fountain.

This town concept can provide housing for an old town, while the rest of the surrounding land could return to the kind of openness and adjacent farms that it probably had originally. Each would benefit the other. The new housing would enjoy the benefits of companionship and connection with the past. In turn it would provide a sound economic foundation for the shops and public spaces of the original town. The main portion of the small town and its historic importance would be preserved with a sound financial basis.

Viewed from beyond the surrounding areas of open space, this new concept town almost entirely blends into the landscape as if a natural hill. This makes it very acceptable as a neighbor in almost any location. The possibilities are limited only by our imaginations and our willingness to be open to new ideas.

Another perspective: our group was asked to study and propose a redevelopment project next to a Southern California mission in 1959. The area was next to downtown where descendants of those who built the mission lived. Many homes were the original adobes with dirt floors. They were generally clean with traditional neat, bare dirt front yards. Scattered between them were new homes of younger, more prosperous descendants, with picket fences, flowers and well-kept lawns. They all had great pride in the Mission and their neighborhood. We recommended the area should remain as it was. This new concept town with its open space would have been a perfect neighbor, giving the city new housing and vitality; but the old area would have been the main attraction.

This new concept town could become part of any historic feature, in a rural area near other small towns and farms, in the middle of a vast suburban area, or, ideally, next to any large city's downtown center. In suburban areas or on large downtown infill sites it will bring with it the qualities and benefits of a natural rural atmosphere. It will significantly improve the value of any existing facilities surrounding or adjacent to it.

# Farms That Gobble Up Suburbia?

*Replaces suburban asphalt and concrete with farms*

Building suburban tracts, swallowing up all that beautiful high-quality farmland, may be one of the smartest things we have ever done.

A couple of hundred years into the future, or sooner, the suburbs may become our best available land for farming combined with this new concept town. By long-term historical standards, the type of structures typically built in the suburbs have not lasted very long. Most of the infrastructure of roads, walks, etc. needs to be replaced about every 50 years.

Meanwhile, the land is protected from concerns about over-production, fertilizer poisoning the soil, and the construction of higher-quality structures that would be too expensive to tear down. Suburbia could protect farmland for a hundred years or more. That could be about the time we will be desperate for more places to grow food.

It's hard to accept suburbia as being a good way to protect farmland for the future. With the other shortcomings of suburbia, that's probably not a good enough reason to build any more than necessary. But such developments will probably remain the mainstay of the housing market for decades.

In the meantime, this new concept town offers an alternative; it's about creating a new kind of relationship between farming and towns. Ag-tourism may even be part of its economic base.

This town's concept efficiently makes farmland an integral part of it. It makes the relationship between town and farms sustainable. Being permanent means housing and farming will be there for the future at a reduced cost. No other housing/town concepts can do that.

Other currently popular town concepts, those that timidly concentrate housing in order to get a little more open space and get people almost close enough to make walking the likely choice, just will not get the job done, especially in the long term. Their typical open space may be a visual benefit, but there are maintenance costs associated with its single use and with all the streets.

This new concept's advantages make it a wiser investment for the buyer and the town. Many other features and benefits were previously described.

Already in the short history of the U.S., entire sections of cheaply built dwellings have been removed for higher and better use. It could be that this new type of farming will be the highest and best use in the future. On a square mile of suburbia this concept gets back over 50% of the land for farming, another 25% for open space, and still has the same number of homes or more per acre.

At the appropriate time in the future, such new types of farms may gobble up worn-out suburbia. Growth limits might encourage the process of replacing suburbia with farmland and new housing.

In areas where aged suburban housing is no longer economically maintainable it can be easily removed and replaced with this new concept town. Where the existing suburban water and sewer lines are in good enough shape to be economically retained, they will be used in the open spaces for supply and drainage. Underground conduits, electrical service, and coaxial cables may find partial reuse. Initially all trees and shrubs would be retained.

Its small footprint allows about 70% of the old housing to remain occupied until the new housing is complete. This concept will provide a variety of housing choices. Enough housing will be suitable for all the economic levels of the old homeowners or tenants. The value of their land alone could provide owners with a good down payment. The flexibility of the new home sites will give everyone a better opportunity to grow their new investment.

If development is done according to a *five-hundred-year* plan, the entire approach can be improved.

New 3-D home sites with the immediately adjacent facilities for a few hundred units could be built in one portion of the ultimate development site. Most of the existing homes that are in good enough shape can remain on the rest of the site. At the end of their useful life, their occupants can move to new 3-D home sites.

The development process can be designed to evolve at the most logical pace. However, a most important *provision*: any improvements to the existing must be understood as temporary. If updated kitchens or baths are needed on the old housing, modular units should be installed that can be reused in future 3-D home-sites. If there are empty lots, any infill should be done with room modules or panel systems that can be reused. This means they can be of the highest quality, so their value may increase like old, high-quality furniture. Any portions certain to be torn down eventually can be maintained in the least costly fashion. This concept makes it easier to make wise investment decisions.

The picture below gives an approximate idea what this new concept might look like located in an existing suburban area. The scale of the two combined images is not exactly the same. (Note: The location shown has no particular significance; this just happened to be a picture that was in my photo file.)

Perhaps certain farmland trusts would be interested in returning 50% of suburbia back into farmland. This could provide the initial investment for this new type of town. This farming, combined with the town, is sustainable for hundreds of years. Perhaps it will be a more attractive alternative than any other choice for the land, … a better development and farming trend.

# Town for New-Growth Areas

*Allows this new life form to blossom*

Wherever it's intended to be built, when people first see the aerial view of this new concept town, they notice all the surrounding open space. At first most think this town must be in a new growth area. They won't immediately realize 70% of the land, which is the surrounding open space, is actually an integral and functional part of the town. That open space is like the new concept town's root system that connects it to the earth. It will be part of the town forever, even if it's built in a downtown area.

This concept can be built on a suitable site in any of the locations described in previous essays. The surrounding open space acts as its

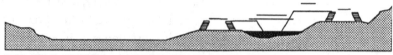

**ONE MILE SECTION .......TOWN ON HILLSIDE STEPS**

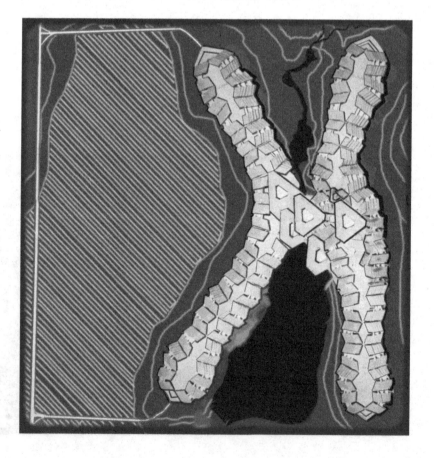

protector, a comfort barrier, which gives it nourishment and insulates it from any negative aspects of the surrounding man-made environment. Wherever it is built, the functional, environmental, and human survivability gives purpose to the protective space wrapped around it.

In each of the previously described locations this concept is influenced by other surrounding man-made factors. In a way of thinking, its innate natural character is compromised in those settings, and that will remain until its influence becomes more understood and powerful, ... as a positive impact. Eventually the power and quality of its own character will overcome those more conventional man-made surroundings, and eventually it will become the primary influence and inspiration for everything around it.

Just imagine, it could revive other cities like those in the L.A. basin with natural open countryside and what appear to be tree-covered hillsides as far as the eye can see.

But in new growth areas where there are no external man-made impacts, it will have an opportunity to test its limits and stretch our imaginations.

It could blend into the slopes along the edge of farming valleys in many parts of the world. Nestled into a small tributary valley, major wings of the town could stair step up the slope on one side and retain the natural farmland.

It would fit particularly well in foothills of the West. Perhaps it could become the modern day "hill towns" of California.

During its initial design stage it can be adjusted and conformed to site conditions. It will, over time, remain adaptable and flexible enough to meet the changing needs of its inhabitants. Over hundreds of years, generations of those living in it will mature, change, and adapt to it, ... and it will adapt to them.

Other than by example, its impact on adjacent human habitats will be minimal; usually a single transit route through the countryside will connect it. But there will be many other connections, communications and interactions of all imaginable forms. There could even be tree-covered country lanes along creeks past groups of three or four farm houses with stands offering seasonal fresh fruit, vegetables, and honey.

This new concept town and everything in it is designed to integrate with nature, designed to eventually get as close as possible to the levels of perfection common in natural life-forms.

When built in a new growth area, this new concept town will discover what it is like to be in an entirely natural habitat. Its inhabitants will discover that they and the entire complex of systems around them will function and be at one with nature.

It is interesting to think about and realize the potential importance of a town that is more efficient than any other town at everything a town is expected to do; as a result, as our wisdom grows, it will probably do more than was ever expected. For while it is made by humans and will be nurtured by them over hundreds of years, as it matures with its enriching layers of life and patina, it will eventually arrive at a balance and harmony with its natural surroundings. The humans that inhabit its spaces may also eventually discover that same balance and harmony, and together they may both discover they are more natural than man-made.

*Part Seven*

# First Prototype?

# Approaches to Affordability
*More than just reducing the price*

Many areas of the United States are struggling with the demand for affordable housing. Quick fixes or temporary support will probably not be adequate for the long term. Reducing cost and making life more economical is not just about providing money or housing. The greatest benefit of this new concept town is simply achieving livability more efficiently, … for everyone.

The flexibility of the home site allows housing to be built to meet small budgets. It also gives occupants the chance to improve their own living condition. The housing can be adjusted to meet changing needs and budgets.

There are many other ways this new concept town contributes to affordability. Efficient use of materials, energy, and space is the fundamental principle, … the arrangement of the entire town, … everything working together in a grand symbiotic relationship makes affordability practical and sustainable.

The following list recalls some of the features mentioned throughout this book that make living here more affordable:
- Flexibility to build only as necessary for budget and needs.
- Making living space smaller when desired.
- Freedom to rent extra space for income.
- Sweat equity opportunity.
- Extended family accommodations.
- Safety enhanced by design of the community and neighborhood.
- Savings on fees and costs of moving.
- Proximity to everything.
- Eliminating the need for two cars.
- Eliminating the need and expense of any car for many.
- Making it practical and easy to walk to everything.
- Efficient transportation for longer trips.
- Proximity of work and family activities.
- Savings in time.
- More time to do everything.
- Preserving agriculture; good long-term economics.
- Recycling of materials and all products.
- Reduced use of expensive ag-chemicals.
- Use of minimum land and resources.
- Long life of buildings leading to less waste.

- Improved reuse, little waste of anything.
- Permanent use saves more than all other options.
- General health improved with more walking.
- Health care access and service enhanced.
- Child care easily shared.
- Senior care more easily shared.
- Seniors and children don't need to be driven places.
- Appreciation of nature and neighborhood happens naturally.
- More opportunity for all ages to contribute.
- More opportunity to be an active part of the community.
- Cost of infrastructure spread over many years.
- Once paid for, free infrastructure for centuries.
- Easily maintained infrastructure.
- Dramatically reduced energy needs.
- Financing advantages for the town and each home.
- Tax advantages.
- Education and sense of community part of every day.
- Stable neighborhoods and friendships.
- Improvement of all aspects of life.
- Livability.

This new concept town offers more ways of making housing and livability affordable for everyone. Features such as these make possible and affordable what money can't buy in any other town concept currently being built.

# Solving General Plan Conflicts

*If not done, worse impacts as centuries pass*

The following comments compare features of this new concept town with proposals in the draft General Plan for Monterey County, California, as it existed in July 2003. Similar comparisons will apply to most places where general plans are based on typical concepts of towns. This new concept's first prototype will be built when people fully appreciate the advantages of a town designed primarily for people.

Our current concepts for housing and towns may *never* allow us to solve the conflicts between people's needs, agricultural needs and responsibility to the natural environment. After fighting over planning issues, spending much time and money, everyone is usually forced to accept less-than-satisfactory compromises. Over decades, it happens again and again.

Without better concepts, conflicts and impacts will only become worse as *centuries* pass.

This entirely New Concept Town (NCT) may actually be able to satisfy *everyone's* objectives. Examples for comparison:

(1) The General Plan intends the rural 3500 acres northeast of the county seat for 62,000 people.

NCT can provide housing for the same population, but also on the same amount of land have 1700 acres for farming plus 800 acres for natural and recreational open space.

(2) The General Plan allows 40-acre parcels in agricultural areas to have three homes plus farm buildings. Along with the roads and driveways, that could use 10% or more of the land.

On only 15% of the land (just 5% more), NCT provides the same number of homes *as if 100% of the land were covered with General Plan housing developments.*

NCT uses the remaining land (85%) for some recreation, but mostly for farming as an integral part of NCT's sustainable self-contained water and waste recycling eco-system.

(3) NCT can turn less-than-desirable land into high-quality farms, potentially adding farmland even if the population is planned to double.

(4) The General Plan cannot make new developments be compact enough for efficient local or regional transit. Most will still need to provide alignments, distances, and spaces based on the needs of cars.

Every NCT home will be within a five-minute walk of NCT's own self-supporting transit system. This transit system's convenience and efficiency will create maximum use and minimum impact on existing surrounding areas or streets.

(5) Rather than typical back yards viewing fences and neighbors' houses, all NCT homes have private back yards overlooking hundreds of acres of open space, like being away on a hillside in the country.

(6) Rather than typical front rooms viewing houses, often empty streets and parked cars, all NCT homes have front rooms and porches overlooking an immediate neighborhood play area, neighbors' porches close enough to wave greetings, plus views of Main Street with shops, sidewalk cafes, parades, and no cars. Interesting activities encourage more front porch use, making for safer neighborhoods and a real feeling of community.

(7) NCT homes can be built to meet any owner's budget and needs. Each home could start as a one-room house for a young couple, grow to a complete four-bedroom house, and empty nesters could later retreat into a one-bedroom portion and rent the rest for retirement income. Total flexibility for all ages, future generations, and extended families encourages wonderful neighborhoods and new approaches for affordability.

(8) NCT reduces all environmental impacts: water, energy and construction material use, for centuries.

(9) NCTs fundamental principle is that towns are for people, and everything is within a short walk. Cars have a convenient but separate place.

The absence of cars will improve livability for homes, neighborhoods, and their special section of Main Street as well as it has for Italy's

Venice or its Hill Towns, as well as it has for our shopping malls where cars parked out of sight below the shops.

(10) NCT is based on the principle that *continual long-term use* is more efficient than recycling or reuse. NCT is designed so at least 75% of each home's initial cost is permanent, protected, and near maintenance-free. Those durable elements of construction are paid for only once but designed to last hundreds of years. That will allow long-term low-cost financing, another approach improving affordability.

Long-term thinking can change the entire way we design and plan. Many of those great places in the world we love to visit are hundreds of (and many over a thousand) years old. A lot of people would be without affordable housing if not for so many very old stone structures. General Plans for 25 or 50 years — that's a very short time. Did anyone plan for Rome to be there a thousand years? Probably not, but it suggests that our towns will make some kind of a long-term mark.

NCT can easily be included as a special section of new General Plans. It may provide an alternative that could resolve current conflicts and remove car-oriented limitations on opportunities for a more imaginative future.

# Getting It Built

*New opportunity for housing newcomers*

The future of any new concept is greatly affected by how it is first introduced to the marketplace. To do it in such a manner that it will be correctly and fully appreciated is a delicate challenge. Presentation of the concept is the first step, but actually building the first complete example, a prototype town, is even more important. Certainly that is the only way it will be possible to fully understand or appreciate something of this scale and comprehensive nature.

Not being done correctly the first time can spoil a good idea. Every step and detail is important. This book provides only an outline, a broad sketch of the idea. Even though it suggests many details, it will take at least another book to fully describe each of the systems that have been developed, as well as a broad-based design team to fully coordinate technical aspects.

This concept offers a significant change to a major public element, similar to when the flying buttress was introduced to cathedrals, creating possibilities for entirely new experiences. Each of the extended neighborhood streets is on the same order of magnitude as those cathedrals, so the size of the concept is not unreasonable; it's easily built with standard construction.

Properly built, this concept can do more for humanity's sustainability than any single project humanity has ever built. New arrangements allow traditional features to provide much higher levels of livability.

Past cultures have built great monuments reflecting important characteristics of their time. In a manner of thinking, this new town concept could be our monument to livability for individuals in their neighborhoods and a demonstration of co-existence with our natural surroundings. In addition to preserving for history how much we value living on Earth, it might demonstrate design patterns helping other towns and countries achieve maximum livability.

This new concept town's first prototype would be perfect as a bedroom village near our most advanced industries, or as a year-round resort or retirement community; with its integral approach for affordability, those who work in it could also live in it.

To complete the next step an administrative and design team fully committed to the concept will be required, a team that is motivated only to find and refine the optimum applications that address all the objectives involved, with no other agenda or outside influence. There will need to be many specialists working together. There will be a role, perhaps performed by a small team, like that of a concert director bringing together all the subtle social, spatial, and technical instruments of design, all dedicated to making this an incredibly beautiful new form of human-created reality.

Bringing this concept to the public is an important next step. Educating the market is a big challenge. A large part of the housing market is already seeking something that combines the best features of a tract house and a condominium. People are also concerned about environmental issues and re-creating that small town character. It is necessary to find the best way to present this concept to buyers.

A very high quality video is needed to make this entirely new and comprehensive alternative to town design easier for more people to visualize, ... possibly a 20-minute PBS special. Once they are better able to see what it would be like to walk through it, I believe many will want to live in it. Some may want to get their names on a list to buy one of its home sites, to make a reservation for an ideal location to build the home of their dreams.

The repetitive design of the primary structure allows for maximum construction and economic efficiencies. Developers and investors can expect excellent returns. When anything new of this magnitude is attempted there are the normal unknowns; otherwise it is basic, straightforward construction. Large infrastructure for housing is a new market. High-quality manufactured modular units would also have a new market. It would be surprising if any conventional house builders would be interested in constructing the infrastructure, but they could build their same systems within this new type of home site.

If the government attempts to do it, the complexities and the required bureaucracy to make sure everything is done correctly to protect everyone's respective interests can have a way of growing bigger than the project itself.

For the first prototype, perhaps the conventional or most obvious approaches may not be the best. Past wisdom not properly guided in this entirely new context could have a negative effect. However, that same wisdom is essential for getting everything right the first time.

Maintaining the purity of the concept is very important. It seems there may only be four types of entities which are best positioned to build the first prototype of this new concept.

First: A large international construction company could have the required world experience and understanding. Perhaps they might recognize this as a significant opportunity. They have more technical skills than the conventional house builder, even though this area of construction may be new to them. The new concept town is of a size that many companies are accustomed to building. They might also recognize markets in other countries.

Second: A foreign government with adequate decision-making authority might recognize the improvement this offers as compared to all the other types of housing they are providing for their people. It could significantly improve lifestyles, retain traditional city life, and give the country views U.S. suburbia imagines it provides. This concept could satisfy the dreams of growing middle classes better than any other choices.

Third: A new concept town could also be built by a large group of people bound together by a strong common interest or stage in life. They might join together in a corporate organization formed to find ways to meet their dreams. This concept offers community features with more potential than any other alternative. A sense of community would already exist even before it was built.

Fourth: Certain nonprofit foundations may be the type of entity best suited to develop the initial demonstration project in the United States. Most are sincerely interested in many of the issues this very comprehensive concept solves. If it proved to be too large for any single foundation to finance, two or more could join together. Properly done, 70% of each homeowner's investment could be a tax deduction. This could make every home less expensive and be a way to provide affordable housing for those who need it.

Whoever builds the first prototype must agree not to change the fundamental concept. Everything about this new concept town, including the process and developer, will be new. Identifying the real client as each individual user was the first challenge. Finding the right individual or group to make it happen is a bigger challenge. Ideally, they may also want to live in it, … be the real client. The future and the opportunity await.

# Appendix

# Bibliography

Alexander, Christopher. *The Timeless Way of Building*. New York: Oxford University Press, 1979.

—. *The Nature of Order, An Essay on the Art of Building and The Nature of the Universe: Book One, The Phenomenon of Life*. New York: Oxford University Press, 2001.

Antoniou, Jim. *Cities Then & Now: Paris, San Francisco, Rome, Prague, Jerusalem, and others*. New York: Macmillan, 1994.

Benyus, Janine M.. *Biomimicry: Innovation Inspired by Nature*, New York: William Morrow and Co., 1997.

Bentley, James; Palmer, Hugh. *The Most Beautiful Villages of Tuscany*. London: Thames and Hudson Ltd., 1995.

Bernick, Michael; Cervero, Robert. *Transit Villages in the 21st Century*, New York: McGraw-Hill, 1997.

Blake, Peter. *God's Own Junyard*. New York: Holt, Rinehart and Winston, 1964.

Calthorpe, Peter. "Transforming Suburbia" from *The First International Ecocity Conference Report*, Berkeley: Urban Ecology, 1990.

—. *The Next American Metropolis: Ecology, Community, and the American Dream*. New York: Princeton Architectural Press, 1993.

Carver, Norman. *Iberian Villages: Portugal & Spain*, Kalamazoo: Documan Press, Ltd., 1981.

—. *Italian Hilltowns*, Kalamazoo: Documan Press, Ltd., 1970–95.

—. *Silent Cities of Mexico and the Maya*, Kalamazoo: Documan Press, Ltd., 1986.

—. *North African Villages:Morocco, Algeria, Tunisia*, Kalamazoo: Documan Press, Ltd., 1989.

—. *Form & Space in Japanese Architecture*, Kalamazoo: Documan Press, Ltd., 1993.

—. *Silent Cities of Mexico and theMaya*, Kalamazoo: Documan Press, Ltd., 1986.

Cotton-Winslow, Margaret. *Environmental Design: The Best of Architecture & Technology*. Glen Cove: PBC International, Inc., 1990.

Doxiadis, Constantinos. *Ekistics: An Introduction to the Science of Human Settlements*. New York: Oxford University Press, 1968.

Duany, Andres and Elizabeth Plater-Zyberk. *Towns and Town-Making Principles*, Alex Krieger and William Lannertz, eds., Cambridge, MA: Harvard Graduate School of Design; New York: Rizzoli, 1991.

Feinginger, Andreas. *The Anatomy of Nature*, New York: Crown Publishers, Inc., 1956.

Fletcher, Sir Banister. *A History of Architecture on the Comparative Method*. New York: Charles Scribner's Sons, 1896, 1954.

Gallion, Arthur; Eisner, Simon. *The Urban Pattern: City Planning and Design*. Princetion: D. Van Nostrand Company, Inc., 1950.

Garreau, Joel. *Edge City: Life on the New Frontier*. New York: Doubleday, 1991.

Girouard, Mark. *Cities & People: A Social and Architectural History*, New haven & London: Yale University Press, 1985.

Ishii, Kazuo. *Membrane Designs and Structures in the World*. Tokyo: Shinkenchiku-sha Co., Ltd., 1999.

Jacobs, Allan. *Great Streets*. Cambridge: MIT Press, 2001.

Jacobs, Jane. *The Death and Life of the Great American Cities*, New York: Random House, 1961.

Jacobs, Jane. *Cities and the Wealth of Nations*, New York: Vintage Books, 1984.

Korn, Larry; Editor. *The Future is Abundant: A Guide to Sustainable Agriculture in the Pacific Northwest*. Seattle: Washington Tilth, 1982.

Kunstler, James Howard. *The Geography of Nowhere: The Rise and Decline of America's Man-made Landscape*, New York: 1993.

—. *Home From Nowhere: Remaking Our Everyday world for the 21st Century*, New York: Touchstone, 1998.

Lennard, Suzanne H. Crowhurst; Lennard, Henry. *Liveable Cities*, New York: Center For Urban Well-Being, 1987.

Lowe, Jeanne. *Cities in a Race with Time; Progress and Poverty in America's Renewing Cities*, New York: Vinatage Books, 1968.

Lynch, Kevin. *The Image of the City*, Cambridge MA: MIT Press, 1960.

—. *Good City Form*, Cambridge MA: MIT Press, 1981.

Marshall, Alex. *How Cities Work: Suburbs, Sprawl and the Roads Not Taken*, Austin: University of Texas Press, 2000.

McHarg, Ian. *Design with Nature*, Garden City: Doubleday / Natural History Press, 1969.

Mollison, Bill. *Permaculture: A Designers Guide*. Seattle: Tagari Publications, 1997.

—. *Introduction to Permaclture*. Seattle: Tagari Publications, 1997.

Muench, David; Pike, Donald. *Anasazi: Ancient People of the Rock*, Palo Alto: American West Publishing Company, 1974.

Munford, Lewis. *The City in History*, New York: Harcourt, Brace and World, 1961.

Neubert, Karel; Royt, Jan. *Treasures From the Past: The Czechoslovak Cultural Heritage*, Prague: Odeon, 1992.

Newman, Oscar. *Defensible Space: Crime Prevention Through Urban Design*, New York: The Macmillan Company, 1972.

Perring, Stefania; Perring, Dominic. *Then & Now: Coloseum, Arcopolis, Pyramid of the Moon, Machu Picchu and others*. New York: Macmillan Publishing Company, 1991.

Read, Samuel. *Leaves from a Sketch-Book: Pencillings of Travel at Home and Abroad*, London: Sampson Low, Marston, Low, and Searle, 1875.

Register, Richard. *Ecocities: Building Cities in Balance With Nature.* Berkeley: Berkeley Hills Books, 2002.

Saarinen, Eliel. *The City: Its Growth, Its Decay, Its Future.* Cambridge: MIT Press, 1943.

Safdie, Moshe. *Beyond Habitat,* Cambridge: MIT Press, 1970.

Schneider, Kenneth. *On the Nature of Cities; toward enduring and creative human environments,* San Francisco: Jossey-Bass Publishers, 1979.

—. *Autokind VS. Mankind: An Analysis of Tyranny, A Proposal For Rebellion, A Plan For Reconstruction.* New York: iUniverse.com, 2001.

Schneider, Kenneth; Zellmer, Gene. *The Community Space Frame: An Integral Approach to Urban Development.* Berkeley: University Extension; City, Regional and Environmental Planning, 1975.

Scully, Vincent. *American Architecture and Urbanism.* New York: Henry Holt and Company, 1988.

Sitte, Camillo. *City Planning According to Artistic Principals,* New York: Random House, 1885, 1965.

Soleri, Paolo. *Arcology: The City in the Image of Man,* Boston: MIT Press, 1969.

—. Omega Seed: An Eschatological Hypothesis. New York: Anchor/ Doubleday, 1981.

Southworth, Michael; Ben-Joseph, Eran, *Streets and the Shaping of Towns and Cities,* New York: McGraw-Hill, 1997.

Tange, Kenzo; Kawazoe, Noboru; Watanabe, Yoshio. *Ise: Prototype of Japanese Architecture.* Tokyo: Zokeisha Publications Ltd., 1965.

Venturi, Robert; Scott-Brown, Denise; and Izenour, Stephen. *Learning from Las Vegas.* Cambridge: MIT Press, 1977.

World Congress on Environmental Design for the New Millennium, *Conference on Green Design Proceedings,* Seoul, Korea: 2000.

Wu, Norbert. Splendors of the Seas. Hong Kong: Hugh Lauter Levin Associates, Ltd, 1994.

# Credits

| Page | | |
|------|---|---|
| 50 | *Image 1* | Street patterns traced from *Great Streets* by Allan B. Jacobs |
| 84 | *Image 1* | Steve Proehl |
| | *Image 2* | Wayne Thom |
| 85 | *Image 1* | Allen Weber |
| 86 | *Image 1* | Wayne Thom |
| 87 | *Image 1* | Wayne Thom |
| 94 | *Image 1* | Department of Housing and Urban Development Photo File |
| 95 | *Image 1* | Kaiser Graphic Arts |
| 110 | *All Images* | Wayne Thom |
| 111 | *All Images* | Wayne Thom |

All other images by L. Gene Zellmer, AIA

# About the Author

Gene Zellmer enjoyed a unique combination of typical experiences that contributed to forming his entirely new concept for towns. He has always believed you could learn something from every person and experience you encountered.

The farm years (ages 5-10), on the last parcel of cultivated farmland northeast of Fresno, California, next to the Sierra foothills, consisted of very modest and simple joys. He has fond memories of working all day in the vineyards with other families, finding fun in racing to pick the most grapes, and sitting under a grape vine carving imaginary toys out of dirt clods. There are memories of taking breaks on hot, 110-degree days by sitting in six inches of cold water in a small irrigation ditch, watching how the shape of the sand bottom changed as water rushed around half-buried legs or was influenced by rocks and little sand piles. Developing patience must have been automatic, simply observing how so many months of hard work could result in the rewards of harvest time. It was a rare opportunity to be part of a fast-disappearing country life style: there were two horses for plowing; the milk cow, chickens, and a garden for food; pet doves, rabbits, cats, and a dog; and, of course, chores. Everything had to be protected from the coyotes, squirrels, and gophers. There were no neighbor kids.

Life had meaning and purpose. Being a contributor to the family's work was obvious: whatever was done, even if small, made a difference. It must have had some influence on his positive thoughts and sense of responsibility. Waiting for some future thing and the joy of simply imagining it was understood as part of the pleasure. It also was a form of escape; years later it evolved into imagining what it would be like to be in a building never-before built.

Those years included a short period in Taylor, Texas living on the edge of town, a kid's walking distance from a grandmother and other kids in the neighborhood. Rainy summer days occurred often enough to keep the crawdads alive in the ditch nearby. It was a ten-minute walk to a creek that could sustain perch and catfish year-round, with pollywogs, frogs, fireflies, snake doctors, caterpillars, cocoons, and butterflies.

Next, were the bicycle years (ages 10-14), in Clovis, California, a small farming community surrounded by cotton, vineyards, and orchards. It was small enough to know everyone; there were different backgrounds, races, and levels of wealth. Everyone was trying to improve everything, and kids could go anywhere on their bikes. Summer jobs were available for kids, from sun-up to sundown if you wanted. It was mostly piece-work. There didn't seem to be any serious troubles in town; there were the occasional drunks, and people tended looked after each other. There must have been some shared level of common sense.

In the high-school years that followed there were no bikes and a little less common sense. Everything started to change; the music was first, but driving seemed more significant. The greatest highlight was, on five occasions, being able to visit the High Sierra and hiking to magnificence remote places. Spending a week or two surrounded by a totally natural environment, untouched by humans, may have been one of his most valuable experiences.

One summer (age 15) he and his father built a house. They started it the day after school let out, and finished the Friday before school started in the fall. They did the concrete, framing, plumbing, wiring, roofing, plastering — everything, all with no power hand tools.

His big opportunity was college and a scholarship. Living on campus during college at the University of Southern California School of Architecture and Fine Arts was a new environmental experience. It was a lot like living in a small town. The five year architectural, structural, landscape and urban planning program opened his thinking to new perspectives on everything, including history and his childhood. It became a wonderfully intense, difficult, and intellectually exciting time, combining life, design, buildings, and living in the L.A.-kind of big city. His university work received several awards, including the highest thesis award.

A year after graduation, he sold everything he owned to spend three months and 9000 miles on a Vespa motor-scooter in Western Europe, experiencing and walking inside of almost every major piece of architecture between Sorrento, Italy and Stockholm, Sweden.

To fulfill his military obligation he joined the Air National Guard, 144th Air Defense Wing, in 1961. Based on his background he received a direct commission in Base Engineering and was a Captain when he was honorably discharged. The organization of a military base was a special, dictated kind of community with unique challenges. It provided another set of perspectives about people, maintenance, operations, and long-term planning.

Working for Robert W. Stevens, AIA, his mentor and friend, gave him the opportunity to design several buildings after college. Zellmer opened his own office in 1963, and his first four projects all won design awards. From then on, he never had to promote his firm.

He attended the Massachusetts Institute of Technology graduate program in architecture and earned a Master's degree. MIT gave him a full scholarship to pursue his proposed research in fabric structural systems. Top-secret clearance gave him access to the military research in fabric structures being done at Natick Laboratories. MIT provided a small additional grant to employ Natick staff to help in his work.

Living in Boston, not needing a car, and walking or riding the transit system everywhere was a great experience. Being on the fifth floor overlooking the Charles River made living in a city acceptable. He felt that with a few refinements, that kind of density could almost be desirable.

After turning down invitations to stay in the East, he returned to waiting clients in his small home town. He believed he could develop an architecture particularly suited to its people in their unique farming and natural surroundings. There, he felt he could find clients more appreciative of nature and open to original work. He realized this would make it easier to avoid being caught up in the popular architectural trends of the large metropolitan areas.

The partnership of Stevens Zellmer Associates was formed, and it designed many award-winning projects between 1968 and 1975. Developers Design Service, Inc., with projects in many different states, was also formed during that period. L. Gene Zellmer Associates was formed in 1975. Zellmer also formed the development company GR.IN. Inc. (Growth Investments).

Beginning with all types of farm work, and then getting a carpenter apprentice license at age 16 and doing carpentry during college summers provided a hands-on understanding of materials. It encouraged a practical confidence focused on inventing anything to make work easier, simplify construction techniques, and control costs. This was exemplified even in his earliest design work. He has been a pioneer in architecture, often years ahead.

Inspired by satellites, space suits, and Frei Otto's work, in 1961 he imagined we would eventually invent permanent fabrics. That would make possible the *first new building system* for permanent buildings since the Romans, over 2000 years ago.

His 1964 and 1965 structural fabric research at MIT resulted, 15 years later, in the world's first applications on a department store, church, residence, and below-grade office building. His work has been published internationally and in *Time* as one of the Five Best Architectural Designs in 1981.

He has designed many types of projects: home, office, school, bank, convention center, civic, shopping center, apartment, HUD senior and low-income housing, hotel, postal, military facility, hospital, legal, dental, and medical projects, plus new community master plans. One of the clients was the largest apartment developer in the United States for many years. Many projects were given AIA Design Awards of Merit, Honor, and Excellence, plus a Builder's Gold Nugget Award. Two projects received the only AIA Enduring Awards for Excellence initially given in his region. There was never enough time, if anything was submitted for publication, someone else must have arranged for it.

The urge to invent resulted in the first applications of silicone rubber to attach and hinge glass, below-grade low-cost low-energy housing and offices, low-cost housing at one-quarter of tract house cost, cooling effects of metal double-roof systems, berm-wall construction, the then tallest (12-story) earthquake zone concrete block housing project, solarium-heated apartments, roof-truss attic room additions, door and wall systems for unskilled labor, and sweat equity housing systems. He formed his own construction development company that built and successfully tested all the ideas too experimental to do on client projects.

His development company built medical and dental office buildings, PUD offices, apartment developments, PUD duplex developments, duplex subdivisions, housing subdivisions and experimental private homes.

He completed his Bachelor of Architecture degree at the University of Southern California in 1960 and his Master of Architecture degree at the Massachusetts Institute of Technology in 1965. He has been licensed in California, Nevada, Florida, Georgia, Texas, and Iowa.

This is a rough outline of events prior to starting his 20-year quest seeking understanding of people's needs in housing and towns. Turning away clients, not submitting his name for any new public projects, and essentially closing his office and moving to a new town seemed to be the only way to get time to do this new research. Even then, several exciting new projects he was not able to turn down occupied most of his time for another ten years. Working at home without the daily office routine and the responsibility to a staff, however, gave him time for study and travel.

Deciding to make this transition was probably the most difficult challenge he has ever faced. It was difficult to talk about this apparent wild dream and its necessity or to explain why he was dropping a promising career to invent an entirely new format for human habitats. He didn't want to say too much because there was always the chance it might be impossible. Even now, with the new concept town presented, many may still not understand its importance. Hopefully that won't take another generation.

Printed in the United States
by Baker & Taylor Publisher Services